Activities for the AL Abacus

Joan A. Cotter, Ph.D.

A Hands-on Approach to Arithmetic

second edition

A Activities for Learning, Inc.

ABOUT THE AUTHOR

Joan A. Cotter Ph.D holds a bachelor's degree in electrical engineering from the University of Wisconsin-Madison, an AMI Montessori diploma for ages 3-6, a master's degree in curriculum and instruction from the University of St. Thomas (formerly College of St. Thomas), and a Ph.D. in mathematics education from the University of Minnesota. Her research was on primary children learning mathematics, especially place value.

She has taught ages 3-6 as a Montessori teacher, taught grades 6-8 as a mathematics teacher, and tutored students with special needs.

Dr. Cotter designed the double-sided AL Abacus and wrote the pre-school through middle-school *RightStart*™ *Mathematics* and *RightStart*™ *Mathematics, second edition* program along with the *RightStart*™ *Tutoring* series. She continues to write and present at national and international conferences.

Printed in the United States of America
August 2021

Copies may be ordered from:
Activities for Learning, Inc.
321 Hill Street
Hazelton, ND 58554-0468

888-272-3291 or 701-782-2000
order@RightStartMath.com
www.RightStartMath.com

ISBN 978-0-9609636-4-5

Contents

Preface

Activities for the Abacus: A Hands-On Approach to Arithmetic is for the elementary teacher and the special education teacher who teaches mathematics. It is written to show how to use the abacus to teach all four operations of arithmetic and some advanced topics.

The book suggests ways to guide the children to make their own discoveries, so that they will feel confident and competent in their own abilities. Frequently, it is suggested that pairs or small groups of children work together.

The AL abacus, which was specially designed for children, has ten rows of beads. Each row has two contrasting colors, which makes quantity recognition possible without counting.

The abacus is used in two different formats. With the wires in the horizontal position, the abacus clearly shows tens and ones. Later, with the abacus turned over and the wires in the vertical position, the abacus shows ones, tens, hundreds, and thousands, with each place value represented by two rows of beads.

Research has shown, that without a doubt, primary children must use their hands in order to learn. A sequence of pictures or even a demonstration is insufficient. Children must have the manipulatives while they work with paper and pencil. The teacher is to show the children how to use the abacus for a particular lesson; the actual learning takes place when the children work similar problems on their own with the abacus.

In contrast, the teacher's editions of most elementary textbooks mention manipulatives, but their suggested use is to teach the lesson, which the children are expected to memorize and apply. They are rarely a part of the children's worksheets.

The companion volume, *Worksheets for the Abacus: Complete Volumes*, provide exercises for the children to do on the abacus. For example, to add, they must actually add, not merely count objects in a picture. Drawings of how to use the abacus appear on each page of a new topic. The worksheets includes addition and subtraction in chapters 1 to 7; multiplication, division, and the remaining topics are in chapters 8 to 11.

The abacus has several advantages over colored rods and blocks. First, the abacuses are easy to store. Second, the beads are arranged like a number line, but in groups of tens. Third, adding two quantities gives an immediate sum with no counting necessary. Fourth, its inherent structure provides a visual model. A five-year-old student named Stan, when asked how he figured out the answer to a problem replied, "I've got the abacus in my mind."

This book introduces the basic concepts of quantity recognition, counting, and reading numbers in Unit 1. Special place value cards help with understanding place value. Next, each of the four operations are approached in three phases: a general introduction, strategies for mastering the facts, and the algorithms. The last two units include the topics of multiples, factors, percentages, other number bases, the Japanese (or Chinese) abacus, and others.

Although children greatly enjoy using the abacus, they spontaneously put it aside once a concept is mastered. They are proud to be able to do their work without it. They also develop the ability to perform mental arithmetic because they understand the structure of numbers.

Today, virtually all fields of human endeavor are using more mathematics than in previous decades. One goal of mathematics education is that the children understand mathematics for appropriate application, not merely be able to perform algorithms. In the near future, less time will be spent on computation. Children using the abacus learn arithmetic in less time, gain a deeper understanding of mathematics, and develop a great love for it.

Notes for Homeschooling Parents

Math Myths

A myth many Americans hold is that the ability to understand mathematics is inborn. On the other hand, Europeans and Asians believe that anyone can learn mathematics with good instruction and hard work.

Another myth is that boys and men are naturally better at mathematics than girls and women. Nonsense. Girls are taught to expect it to be harder for them. Often, they are not encouraged; teachers in schools frequently show them the solution, instead of helping them figure it out.

Today over two-thirds of careers require advanced math. Industry and post secondary schools of all types spend billions every year teaching remedial math. Workers must be able to approach problems confidently and recognize mathematical applications in various situations.

Early Concepts

Here are four practices that will help your preschooler get a good start in math.

1. We know that five-month-old babies can distinguish between 1, 2, and 3 objects. Half of twelve-month-old babies can discern up to 4 objects. Therefore, never count (by pointing to each object) fewer than 5 objects. When you point to objects for counting, the child loses the idea of the whole and assumes you are naming the objects. That is why when you ask a young child after counting 4 objects to show you the 4, the child will point only to the 4th object.

2. English is inconsistent in naming numbers 11-99. The teens in English are especially confusing. Asian children learn to count as follows: 1 to 10, then ten 1 (11), ten 2 (12), ten 3 (13). The twenties are 2-ten, 2-ten 1, 2-ten 2, and so forth. Research shows this helps children understand place value. Therefore, teach your young child to count this way up to 99.

3. We need to give children names for quantities. A good way to do this is through fingers. Most children can show the correct number of fingers as high as their age. Use this same idea, but start with the left hand (because we read from left to right) and continue up to 10. Keep in mind this is not counting.

For example, ask them to show 7, which is 5 fingers on the left hand and 2 on the right hand. It does not matter which fingers to use on a hand. Also, do the inverse: show a quantity on your fingers and ask your child to tell you how much it is. To teach 6-10, use the following rhyme:

Yellow is the Sun.

Yellow is the sun.
Six is five and one.

Why is the sky so blue?
Seven is five and two.

Salty is the sea.
Eight is five and three.

Hear the thunder roar.
Nine is five and four.

Ducks will swim and dive.
Ten is five and five.

4. When the fingers are mastered, use tally sticks for representing quantities. Tally sticks are simply Popsicle or craft sticks. Show your child how to lay out 3, as shown below. Lay them about an inch apart.

Representing 3 with tally sticks.

When your child can both lay out 4 and identify it, show 5 with the 5th stick laid horizontal across the 4 as shown below.

Representing 5 with tally sticks.

Representing 8 with tally sticks.

Continue through to 10; however, to go past 10, start another row. Give your child simple problems to solve with these tally sticks, "What is 4 apples and 3 more apples?" With the 7 sticks laid out, ask, "How can you tell me how much this is without counting?" The solution is to make a 5 with one of the sticks.

Grouping 4 apples and 3 more apples to get 7.

Visualization vs. Counting

In the U.S. counting is considered the basis of arithmetic; children engage in various counting strategies: counting all, counting on, and counting back. Japanese children, on the other hand, are discouraged from counting; they are taught to recognize and visualize quantities in groups of fives and tens.They consider counting to be slow, unreliable, and oblivious to place value.

To understand the importance of visualization, try to see mentally 8 apples in a line without any grouping–virtually impossible. Next try to see 5 of those apples as red and 3 as green; now they become visualizable for virtually everyone.

Because our brains cannot visualize more than 5, it is vitally important to group in 5s. The Romans knew this and used a V for writing their numerals. Originally their 4 was IIII, but they knew they could not continue with straight lines because they would not be able to quickly recognize the numeral. Piano music is written with ten lines, but no one could read it without the space between the two groups of five lines.

Other Pointers

1. This ungraded program is designed for kindergarten through fourth grade. Most homeschooling families spend about 20-25 minutes a day on math. It is impossible and unnecessary to say exactly where a child should be at a particular age. Not only do expectations differ between states, there are variations in local school districts. Children who start this program in kindergarten or first grade do well on standardized tests.

2. It is extremely important that the child be able to enter and name all the quantities from 1-10 without counting before continuing with more advanced concepts.

3. The games are an integral part of the program. Although the *Activities for the Abacus* manual does not specify which game, each chapter of the game book starts at a basic level and gradually increases in difficulty.

4. When your child understands numbers to 100, begin playing the Skip Counting Concentration game (P2). Skip counting is a great help for learning the multiplication facts.

5. Understanding is key in mathematics. Ask your child how she arrived at an answer.

Money and Telling Time

When your child has mastered counting by 5s to 100, it is time to start the money and clock games in the *Math Card Games* book.

Fractions

Fractions are found in the *Math Card Games* book. The fraction manipulative is the linear fraction chart. Be sure to include fractions at every grade level from kindergarten on.

Memorizing the Facts

Unfortunately, many people equate studying math with memorizing hundreds of "facts" and countless rules. To memorize the facts, the child should have a good strategy for determining the answer and interesting practice. Strategies are given throughout the manual and game book. Games provide the practice.

It is another myth that a person learns facts by merely associating a third number with two numbers. For example, if you see 8 and 7, you think 15. Brain research now tells us our brains do not work well that way. Rather, it is more natural to use a strategy. In this case a person might take 2 from the 7, combine it with the 8 and change it to 10 and 5, or 15.

Drilling with flash cards can have detrimental effects. The only person who enjoys flash cards are those who do not need them. Timed tests often place children under unnecessary stress. Brain research shows that a child under stress stops learning. Children may also develop the opinion that they are not expected to think in mathematics.

Later on, after your child has had ample time to practice a strategy, you might ask him to time himself over a period of days and graph the results.

Drawing Board Geometry

This type of geometry usually can be started in second grade. Parents with no previous drawing board experience can become proficient in about an hour.

Attitude

A parent's positive attitude toward mathematics helps determines a child's success.

Mathematics is a magnificent tool for understanding the universe, the atom, and the cell, as well as much in between.

Unit 1
Basic concepts

Before proceeding in formal instruction in arithmetic, children need to master four initial concepts: 1) recognizing quantity, 2) memorizing the counting sequence, 3) one-to-one correspondence, and 4) the writing and reading of numbers. Children do not master these concepts in any particular order and they often work on several simultaneously in their everyday lives.

In this unit, number work begins with helping the children recognize quantities on their fingers and proceeds to recognizing quantities on the AL abacus. Counting to 10 is one of the few activities in mathematics that requires sheer rote memorization. One-to-one correspondence needs coordination between mouth and hand while writing numbers demands hand to eye coordination. Also covered in this unit is even and odd. Finally, the idea of a middle is taught.

NAMING QUANTITIES 1-10

Children often are taught to respond with fingers when asked how old they are. They understand that the more fingers they can put up, the "bigger" they are. This idea will be continued up to 10.

Using fingers for 1 to 5
Hold up your fist so the children can see. Raise one finger and say, **This is 1; can you show 1?** Close your hand and then raise 2 fingers simultaneously. **This is 2; can you show 2?**

Note that no counting is done. Continue with 3, 4, and 5. Be sure that the children use only one hand to show these quantities. If the entire concept is new to the children, teach no more than two numbers at a time.

As a variation say, **Put your hand behind your back and raise 2 fingers. Now look and see if you are right.**

When the children know several quantities, choose one, raise that many fingers and ask, **How much is this?**

Have the children take turns with a partner, showing and naming quantities. They might also ask the entire group to show a certain quantity.

Using fingers for 6 to 10
Show 6 by holding up 5 fingers on one hand and 1 finger on the other hand. **This is called 6. Can you show 6?** Any number over 5 must be shown with all five fingers of one hand plus the extra amount shown on the second hand. For example, 6 may not be shown with three fingers on each hand.

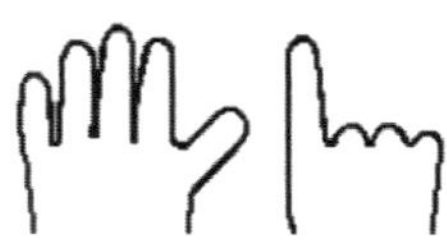

Teach the numbers 7 to 10. Review frequently.

Quantities on the abacus

Help the children correctly orient their abacuses. The abacus needs to be placed flat on a firm surface with the gold dot (the AL logo) in the upper right hand corner. Show them how to clear the abacus by lifting the left edge so the beads fall to the right and then setting it down quietly.

Entering quantities on the abacus is similar to showing quantities on fingers. Each hand is represented by a color. Raise 1 finger and ask, **How much is this? Can you raise 1 finger? Here is how to enter 1 on your abacus.** Slide 1 bead on the top wire to the left edge.

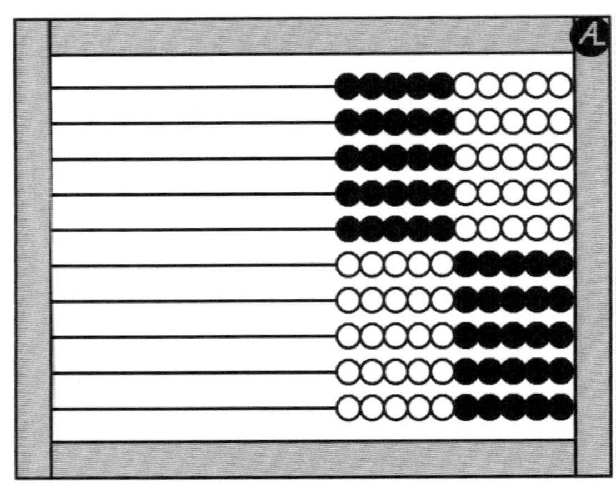

1: Ask the children to say and enter another 1 on the next wire, the bottom wire, or all the wires. Then ask them to clear their abacuses.

2, 3: Now ask the children to show 2 with their hands. Enter 2 on the abacus by sliding 2 beads to the left edge as a unit, not as 2 separate beads. Ask the children to say "2" and enter it. The beads are not to be counted at this stage. Indeed, a great advantage of the abacus over individual counters is being able to name quantities without the tedious counting.

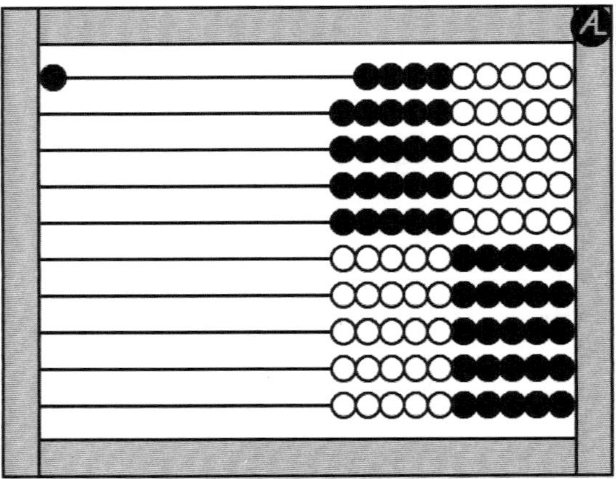

5, 4: Teach 5 before 4, pointing out that 5 is all the fingers on a hand. Show them how to enter 5 as all the beads of one color. To teach 4, ask the children how many fingers stay down when 4 is raised. [1] Thus, to enter 4 means one bead of a color group stays behind.

6, 7, 8: To teach 6, show them that it takes one hand and one finger of the other hand. Ask the children to raise 6 fingers and then enter it on the abacus. Teach 7 and 8 the same way.

10, 9: Teach 10 and 9 in the same way as 5 and 4.

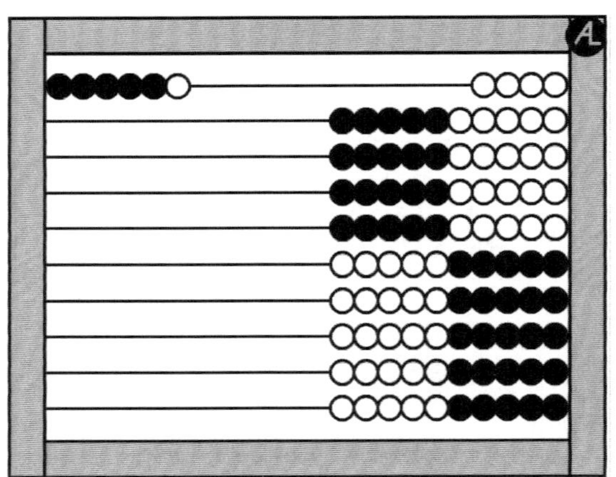

Without saying the quantity, raise a number of fingers and ask the children to enter that amount on their abacuses.

Put a quantity on the abacus and ask a child to name it or express it with his or her fingers. Have the children work with partners, entering and reading quantities.

Matching hands and abacuses

Matching exercises for children have several advantages over drawing lines on workbook pages. The children can be more active, which is essential for learning at that age. Mistakes can easily be corrected and the children can repeat the exercises many times. The children learn how to organize their work. Also, matching can be made into games involving more than one child.

Photocopy, on heavier colored paper if possible, the pictures of the hands and the abacuses found in the appendix and cut them apart. The sets will be easier to separate if they are copied on different colored paper.

Ask one or two children to lay out in a row on a table or
rug the hand cards. The abacus cards are placed face
down on a pile. The children take turns picking up the top
abacus card and placing it below the corresponding hand
card. In general, the more familiar objects should be laid
out in the row and the less familiar stacked and matched to
the row.

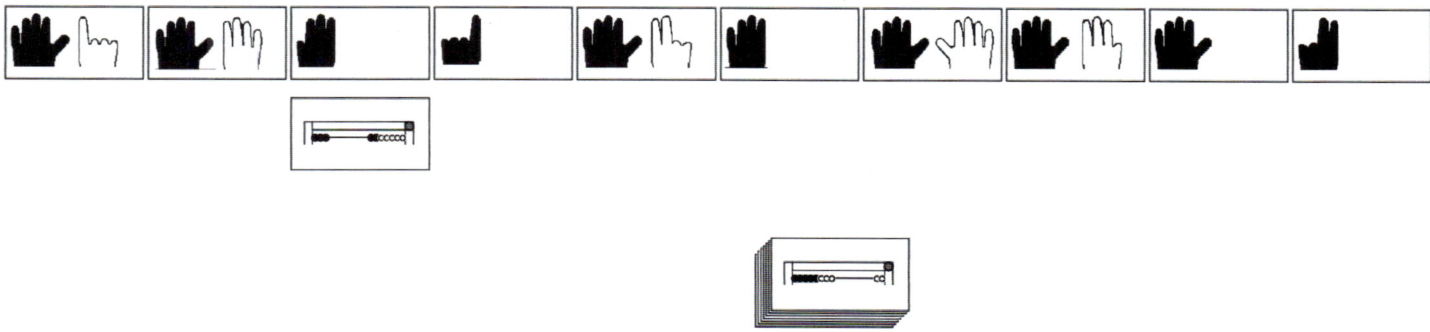

Continue until all the pictures are matched. Show the children how to pick up the pictures in random order after the
game.

Sorting

Spread out the abacus cards. First the child enters 1 on the
abacus and looks for the card of 1 and starts a new row
with it. Next the child enters another bead on the abacus
and looks for that card. It is placed next to the 1. The child
continues until all the cards are found.

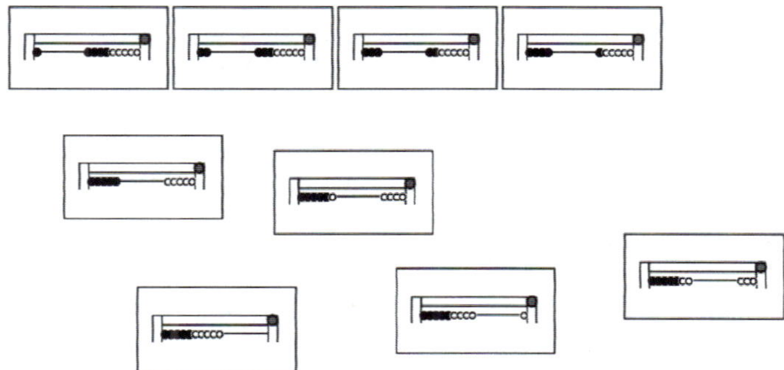

The same activity can be done with the hand cards.

Worksheets

See *Worksheets for the Abacus Volume 1* by the author
for worksheets (1-1 and 1-2) involving matching hands
and abacuses. They may be used for checking mastery.

MEMORIZING 1-10

Memorizing the counting sequence can be started at the same time as introducing quantities.

Memorizing the words

To help children memorize the words and later acquire counting skills, have them repeat after you while you say the numbers slowly, distinctly, and rhythmically. Note that the objective is to learn the words, not to count items.

Teach the following nursery rhyme, which will be also useful later for counting by twos:

> One, two, tie my shoe,
> Three, four, shut the door,
> Five, six, pick up sticks,
> Seven, eight, lay them straight,
> Nine, ten, big fat hen.

During the day use small snatches of time that are frequently wasted to practice counting aloud.

Reading quantities on the abacus

The following is a readiness activity for counting objects. Review by asking the children, **Clear and enter 3 on the top wire. Enter 5 on the next wire; enter 8 on the next wire.** Continue until five quantities are entered. Now start at the top wire and tell the children to touch each wire and name the quantity in unison. Call upon individual children to name them. Later expand the number of quantities entered and read.

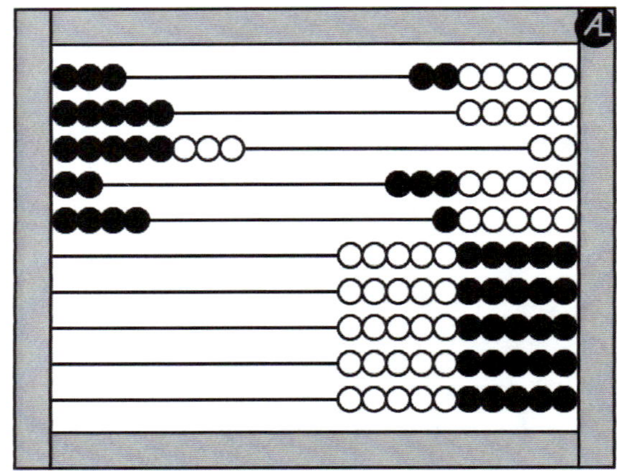

The stairs

Building the stairs is an activity children love to do. Show them the stairs. If they need help in building it, tell them, **Enter 1 on the first wire, 2 on the next wire**, and so forth.

Another way to build the stairs is as follows: clear the abacus and enter 1 on the first wire. On that same wire touch the first bead of the right group. Drop down a wire and touch the bead immediately below it. Now slide your finger and the beads to the left edge. Continue for the other rows.

If they have not discovered the stairs previously, give them time to enjoy it. Then ask them to touch each wire in order from top to bottom while reciting the quantity. Coordinating the touching with the words is an important step necessary for counting.

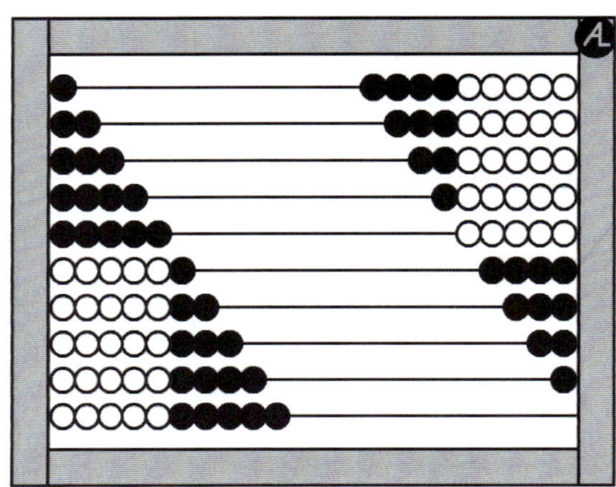

With the stairs the children can practice the counting sequence independently. Later they can practice counting from 10 to 1.

COUNTING

Learning to count objects, also referred to as one-to-one correspondence, can be a difficult skill to master, requiring hand and voice coordination. The task is simpler if the objects are movable and are transported in an orderly fashion. The abacus allows both.

There are two ways to count on the abacus. The first consists in counting all of a quantity. The second is to count out a desired amount from a larger quantity. Both skills are needed for adding.

Counting a known quantity

1. Start with 5 entered on the abacus; demonstrate moving it away from the left edge about 3 cm (1 inch). Show the children how to count by moving each bead in turn to the left edge while rhythmically saying the correct number. **How many beads did we count?** [5] Give the children other quantities to enter and count.

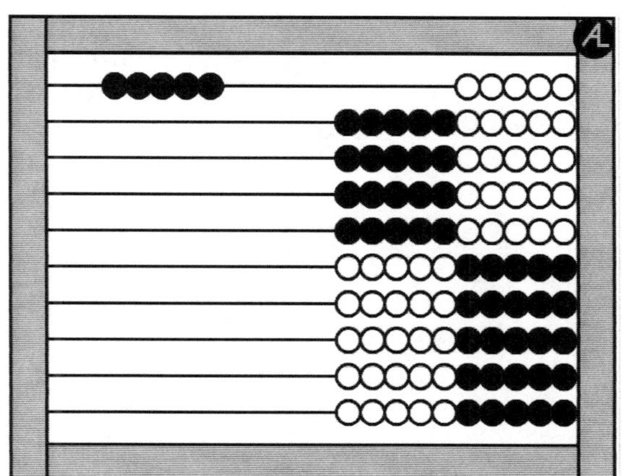

Note that even though a child has mastered the counting skill, it does not mean he or she has acquired the concept of quantity. For example, the quantity 4 includes 1, 2, and 3; whereas, when we say a, b, c, d; d does not include a, b, and c.

2. Tell the children to make the stairs (see the previous page). Next ask them to count down 4 wires, **How many beads are on that wire?** [4] Repeat at random for other numbers.

Give the children interesting objects to count, first objects that they can move as they count and later objects they can only touch, such as in pictures. Fingers are also countable, especially up to 5. Very young children love to count, but they are more interested in the process of counting than in the final result.

Counting to a number

The second step is to count to a certain number and stop. The children must keep in mind the final number while counting. **This time I am going to count 8 beads; you tell me when to stop.** Start with the abacus cleared; rhythmically count each bead as it is moved to the left, letting the children tell you when to stop. **Does it look like 8? Let's see if you can count 4 beads and stop.** Give them other quantities to count.

LEARNING NUMERALS

Children under six have a very strong sense of touch. They need to touch an object in order to understand it, whereas adults usually are satisfied by looking.

Sandpaper numbers utilize the kinesthetic approach and help children in the correct formation of the numbers. The numbers are cut from sandpaper and mounted on a smooth background of a contrasting color. With these numbers, the children learn by feeling, seeing, and hearing the number.

To make a set, trace the set of backward numbers found in the appendix onto the back of medium-textured sandpaper or 3M's "Safety Tread." Cut them out, glue or remove the protective backing, and apply to precut 3-1/4 in x 3-3/4 in (8 cm x 10 cm) Masonite boards, plastic, or heavy cardboard. Place a red dot to indicate where to begin and draw a thin line under it.

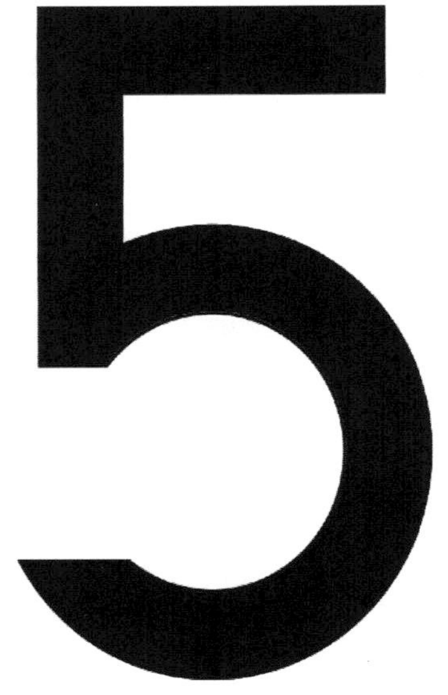

Using sandpaper numbers

Teach one number at a time, starting with 1. Continue through to 9 and end with 0. With your hand held extended and relaxed, use your first two fingers and start at the dot; lightly trace the number in the same direction as it is written, as shown by the dots in the figure. At the same time say, **This is 1; what is this?** [1] Encourage the children to trace it many times while saying it. Review frequently.

Note that the 4 has two starting points. The number 5 is easier to write and often more legible if made in a single stroke. For this reason the single-stroke 5 is preferred. The number 8 is easier to make if the straight line, rather than the "S," is drawn first.

Writing the numbers

Writing numbers is a difficult task for most children; both coordination and memory of the number shapes are needed. Some children are helped gaining control of the writing instrument by tracing simple objects; on a clipboard place a piece of tracing paper over the object to be traced. A tray with baking soda or sand is good for tracing the numbers. Chalkboards are great because the children can easily erase their mistakes.

Listed below are the numbers in the order of increasing difficulty. Also given are phrases that have helped some children to remember how to write them. Without lifting the pencil or chalk (except for 4 and possibly 5), the writer is to pause after each action. This prevents rounded corners and other problems.

    1: Down.
    7: Over and down.
    4: Down, over, and make a 1.

2: Around, then straight.
5 (one-stroke): Over, down, and around.
5 (two-stroke): Down, around, and the top.
3: Around, and around.
0: Around.
9: Around and back to the dot, then make a 1.
8: Down (like 7), around (like 6), and finish.
6: Around and then a smaller circle.

Give the children a page with numbers for tracing (1-9) or writing (1-10).

Practice with the numbers

1. Use a set of cards from 1 to 10 (see the appendix). Hold up a card and ask a child to put that many beads on the abacus. Continue as long as the children remain interested. Repeat on subsequent days.

2. Give each child a set of the numbers from 1 to 10. Ask the children to spread out their numbers so they can see them. Say a number or enter a quantity of beads on the abacus and ask the children to find the card. As each card is picked up, it is set aside in a row or column. Continue until all the cards are picked up.

3. Two sets of cards 1 to 10 are needed for this game for two players; one set is face down on a pile and a second set scattered face up. The first player turns over a card from the pile without the second player seeing it. The first player then puts that quantity of beads on the abacus. The second player looks at the abacus and finds the corresponding card. The two players then compare cards and set them aside. For the next turn, the players reverse roles. The game is over when the pile is gone.

Four simple card games

1. MIXED-UP CARDS. Lay out two sets of cards 1-10 in a row. To put the cards in order, the children may need to use the number strip, which also can be found in the appendix.

While one child turns her back, the other child mixes up two cards in the second row. The first child finds them and corrects it. Then the roles are reversed. Eventually the top row can be removed.

2. MYSTERY CARD. A row of cards 1-10 is laid out in order. The first player turns a card face down without the second player looking. The second player determines what it is and turns it over to check.

3. MISSING CARD. A card is removed and the gap closed. The second child determines what it is and where it belongs.

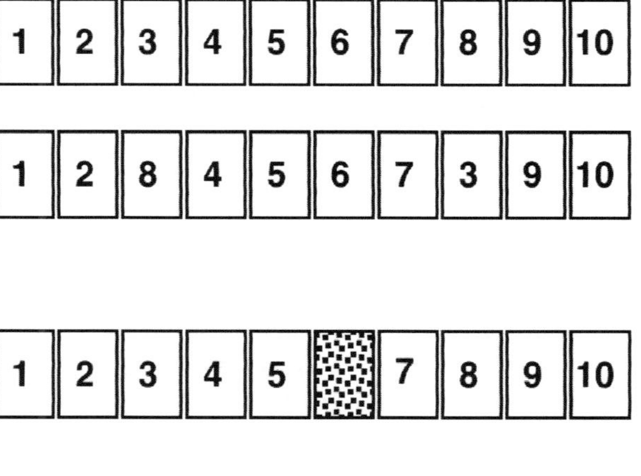

4. GUESS WHAT'S NEXT. The objective of this game is to enable the children to practice saying what number follows another without resorting to counting from one. However, they can still count when they need to. Practice will decrease the response time.

The ten number cards are arranged in order face down in a row. The first child turns any card over and the second child guesses what card comes next, which is checked by turning over the card on the right. Roles are reversed after each turn.

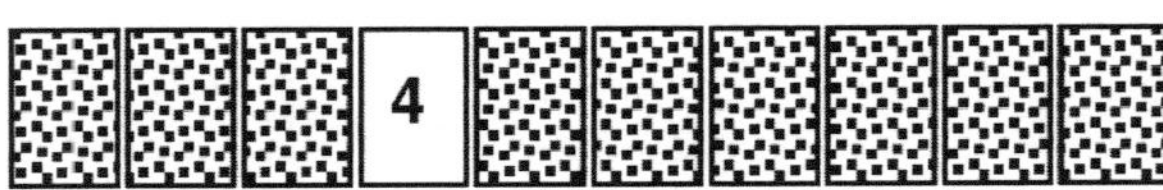

Introducing zero

Show the children a card with 0. Tell them that it is zero and when it is by itself it means none, no beads at all.

Dr. Maria Montessori developed the following two games to help children appreciate the zero. With the children seated in a group, call on a child to clap three times. Ask another child to jump up and down two times. Finally ask an outgoing child to say banana zero times. Gently urge the child to do it while the group supports the child's staying silent.

For the second game, prepare folded slips of paper, one for each player, each with a different number, but include 0. Fold the slips in half so the numbers cannot be seen and place them in a basket. In a second basket place small identical objects; the quantity is equal to the sum of the number on the slips. The children each take a slip in turn, return to their places, and read their numbers secretly. One by one each child takes as many objects as the number on the slip of paper. The child receiving zero, of course, cannot take anything. Check each child's number and quantity. Play the game at least twice.

MATCHING AND MEMORY GAMES

These matching activities and Memory games provide hands-on work without the use of paper and pencil.

Matching with numbers

The matching and sorting activities can now be done with the numbers.

1. Lay out either the hand or the abacus cards in random order; leave a space and lay out the numbers in a row in order below it. Find for each hand or abacus card the corresponding number.

2. Reverse the order of the two rows; put the numbers in order in the top row and the hands or abacuses in random order in the second row.

3. This is the same as 1) above except the numbers are also mixed up.

4. This is the same as 2) above but the numbers are mixed up.

Matching Memory

There are two Memory (sometimes called Concentration) games. In the more traditional game, two cards are turned over; if they match, they are collected and the player receives another turn. In the other game, the cards must be collected in order and only one card is turned over.

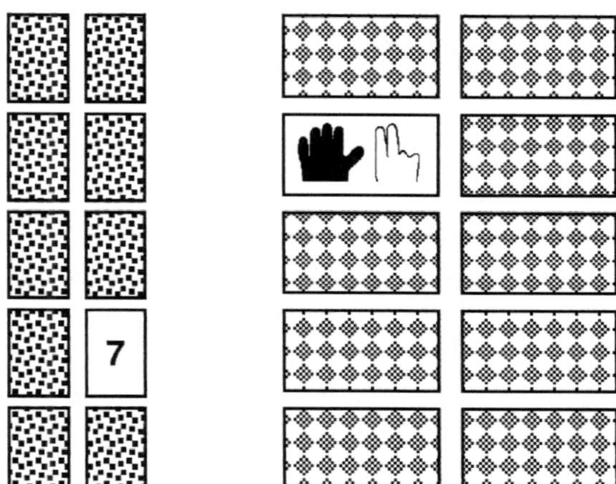

For example, to play Matching Memory with the hands and the numbers, shuffle the pictures of the hands and shuffle the numbers. Lay out each set neatly in two rows. The first player turns over a card, says what it is, and turns over the card thought to be its match. If found, the player collects the cards and receives another turn. If they do not match, the cards are returned face down in the same place and the next player takes a turn.

In-Order Memory

The same cards are used for this Memory game, but the cards must be collected in order. One player will collect the hand cards and the other player, the number cards. The sets of cards may either be laid out separately or mixed together.

To start, the player collecting the hand cards turns over a card. If it is the 1, she collects it and receives another turn to look for the 2. If it is not the 1, she returns it and the other player takes a turn. The player who collects all her cards is the winner.

If desired, give the children various matching activities (1-3 to 1-8) to be done by drawing lines.

EVENS AND ODDS

Primary math programs seldom spend enough time on the concept of even and odd. It has wide application in everyday life and is important in number theory, algebra, and higher math.

The even-odd patterns are far more valuable for children to recognize than the patterns on a die. Adding 1 to the 5-die pattern does not result in the 6 pattern, whereas adding 1 to the 5 even-odd pattern does result in the 6 pattern.

Counting by twos

Use the rhyme, "One, Two, Buckle my Shoe" and teach
the following finger motions: with the palms facing in-
ward and the hands in a fist, count by extending the fingers horizontally, starting with the left thumb. Next challenge the children to say only the numbers, not the words.

Also work on counting by 2s by whispering 1, saying 2,
whispering 3, saying 4, and so forth. Divide the class into
two groups and direct the two groups to count alternately
1 to 10. That is, the first group says 1; the second group
says 2; the first group says 3; and so forth to 10.

Counting the dots

Copy and cut apart the dot cards found in the appendix.
Make one set for every two children.

1. Demonstrate counting the 8-dot card. Starting at the
upper left, count each row from left to right and count the
rows from top to bottom. Call on the children to count the
other dot cards.

2. Have the children work in pairs matching a set of dot
cards to a set of number cards.

3. They can also play the Matching Memory game with
the two sets of cards. Keep the sets separate, and turn the
dot card over first.

4. Give worksheets (1-11 to 1-14) for matching.

Even or odd

Prepare ten tags 2 in x 1/2 in (2 cm x 5 cm) as labels, five
for the evens are to be in a darker dominant color while
the five for the odds are a lighter color.

1. Hold up the pattern card for 8. Show the children how
to determine if it is EVEN: with two fingers of a hand simultaneously touch both dots in the top row. Continue in
each row touching both dots. After the last row say, **This
number is even**. Place an "even" tag under it. Repeat
for patterns 6 and 10.

Next check pattern 7. When the bottom row is reached,
point out that there is a dot for only one finger. **This
number is ODD**. Place an "odd" tag under it. Continue
with other dot cards. Have the children work in pairs assigning the correct tag to each of their dot cards.

2. Ask the children to sort the dot cards into evens and
odds. Then ask them to put each group in order and to
read the results, **Does it remind you of something?**
[Counting by 2s]

3. Now the children can play a Matching Memory with a
set of dots and a set of numbers. Lay the two sets out separately. One child collects the evens while the other collects the odds. The dot card is turned over first. The first
person to collect five pairs is the winner.

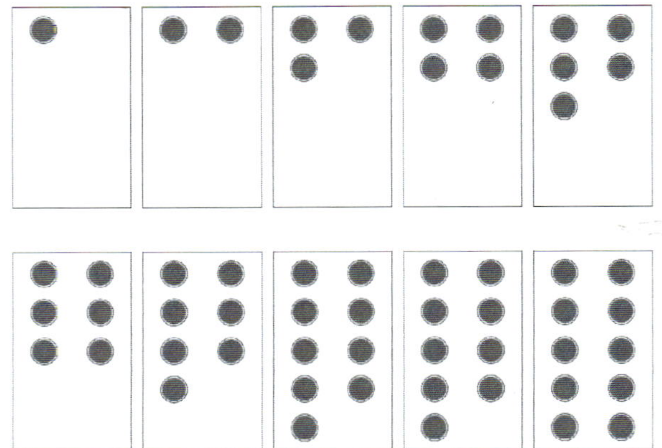

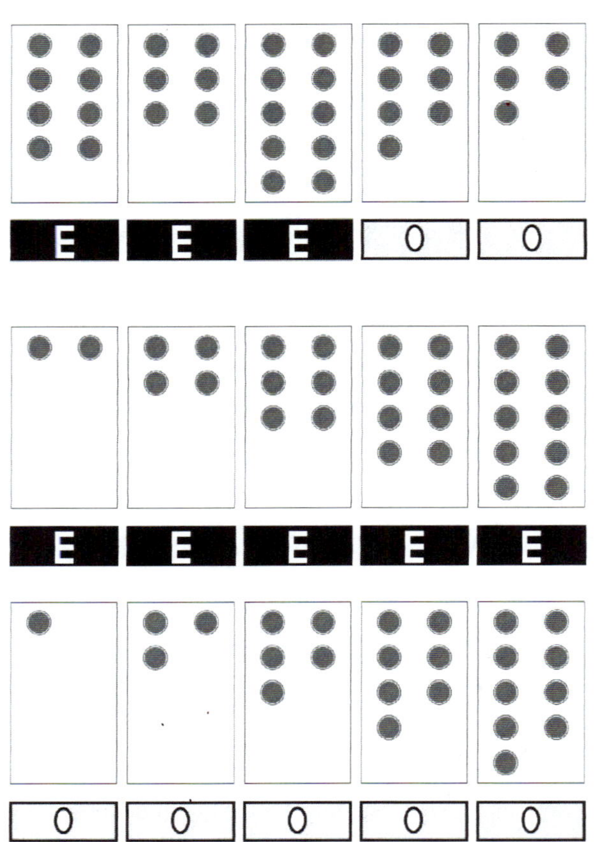

Evens and odds on the abacus

Turn the abacus so the wires are vertical. The logo should be in the lower right hand corner. The right two wires will be used to represent a number. Clear the abacus by lifting the top so the beads fall to the bottom. This format will be used extensively later for representing 4-place numbers.

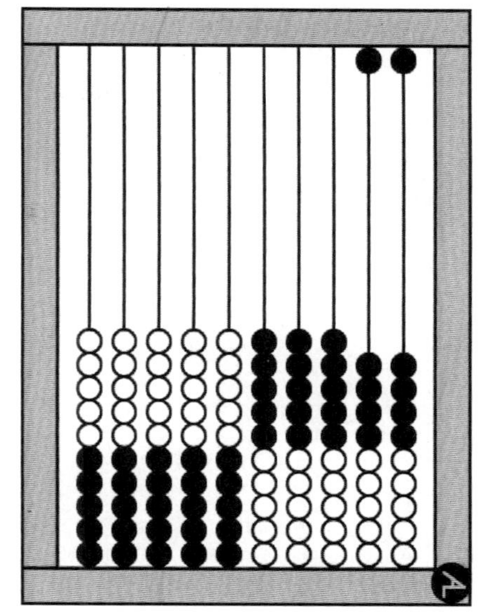

1. Teach the quantities in this new format. Enter one bead on both the ninth and tenth wires. **How much is this?** [2] **Enter 2 on your abacus.** Teach the evens first. After entering 10, ask, **What is special about entering 10?** [All the beads with the darker color] Then teach the odd numbers 1, 3, 5, 7, and 9.

2. After they know most of the quantities, hold up a dot card and ask the children to enter it on their abacuses and then to name it.

3. PUTTING THE CARDS IN ORDER. Ask the children with their partners to spread out the dot cards so they are all visible. One child enters 1 bead on the 9th wire and the second child finds the card and starts the row. The first child enters another bead, this time on the 10th wire. The second child again finds the card and adds it to the row. They continue to 10. The children repeat the exercise with the roles changed.

2. ROW BINGO. This and the following game reinforce naming the quantities and introduce the terms ROW and COLUMN.

Ask the children to put nine dot cards in three rows of three cards each. The leader has a shuffled deck of number cards and holds up one card at a time. Any player having this quantity turns face down his corresponding card. The first player with all cards in any row face down is the winner. The winner may be the next leader.

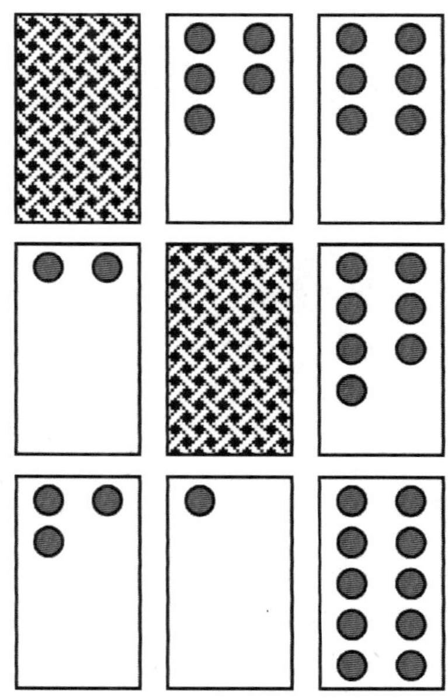

3. COLUMN BINGO. This is the same game, except that the winner is the one who first turns face down an entire column.

4. Ask the children to put their number cards in order. Then ask them to put the corresponding dot card and tag under each number. To help them discover that the tags alternate, move the evens up a bit. Ask a child to read the even numbers and another child to read the odd numbers.

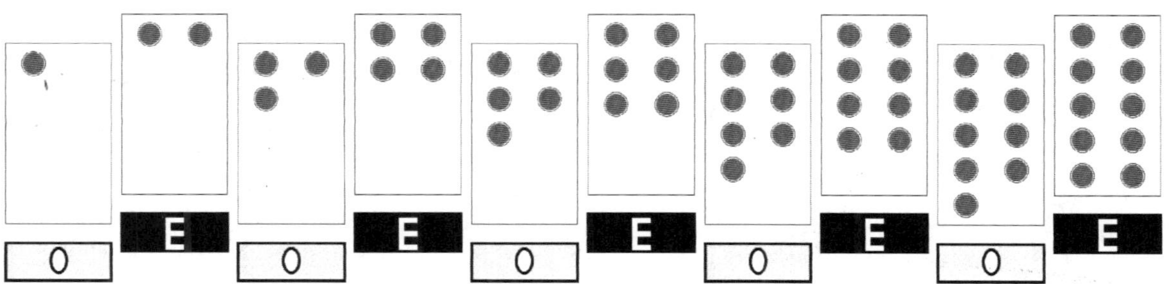

Counting beads by twos

This activity will help the children experience that counting by 2s is really counting with groups of 2. With the wires of the abacus horizontal, show counting by 2s. Move over 2 beads at a time while saying 2, 4, 6, 8, 10.

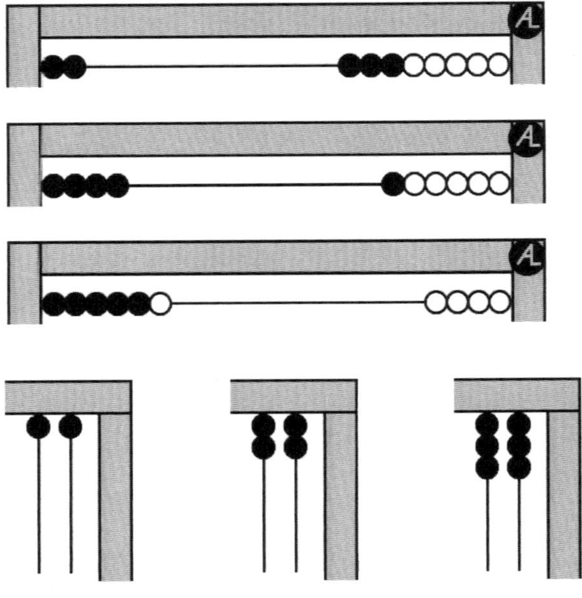

With the wires vertical, count by 2s. Move 2 beads up at a time. This can be accomplished with either the thumb or 2 fingers.

Counting with odds

With the wires horizontal, start with 1, move over 2 beads at a time and say the odd numbers: 1, 3, 5, 7, 9.

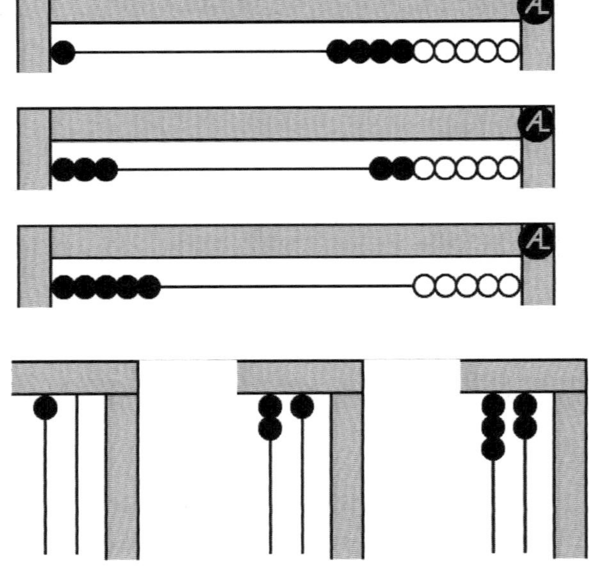

Repeat the activity with the wires vertical. Start with 1 bead; move up 2 beads at a time and count.

The middle

This activity will help the children discover that the odd numbers have a MIDDLE. Enter 9 on your abacus but move it away from the left edge. Simultaneously separate the outer two beads using both hands. Repeat until one bead remains, **This bead is the middle. So 9 has a middle.**

Repeat the activity for the number 8. **Does 8 have a middle?** [no]

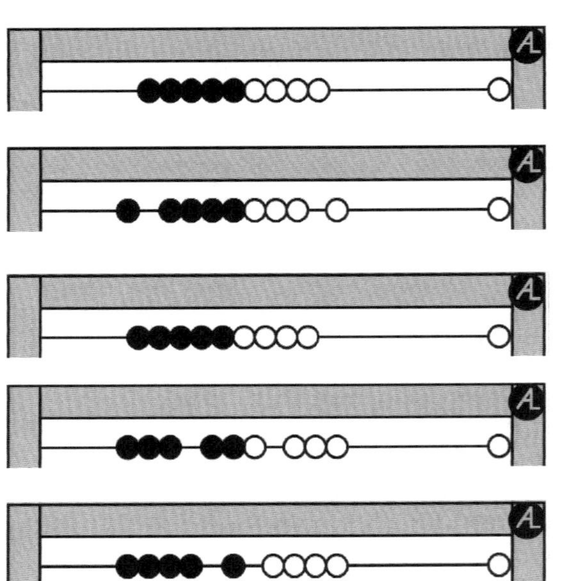

When the children understand, let them work in small groups. Give each group a set of numbers and an abacus. Ask them to sort the numbers according to whether or not it has a middle. When the groups are finished, **Look at the numbers that have a middle; what is special about them?** [They are odd.]

THOUGHT QUESTION: Why do only the odds have a middle? [The one that is left over is the middle.]

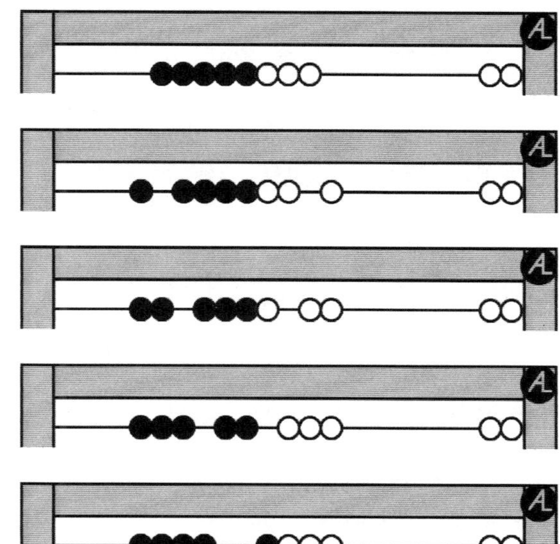

THE ORDINAL NUMBERS

Most children are familiar with the terms: *first, second,* and *third.* The object is to teach the remaining terms to tenth and how to write them. Writing them may be postponed if the children are not completely comfortable with writing the numbers 1-10.

Teaching these names before the tens and teen names will be helpful with them because of the similarity with third, thirty, and thirteen. Likewise it will help with fifth, fifty, and fifteen.

Ordinal numbers in general

Ask six children to stand in a row. Explain that we usually think of a row as beginning at the left. **(Name) is the FIRST person in the row; (Name) is the SECOND person.** Continue to the end of the row.

Ask the children to recite with you or after you the ordinal numbers from first to tenth.

Set up a row of ten different objects with a space between the fifth and sixth objects. Ask various children what is first, third, and so on. Next do the reverse by naming an object and asking, **Where in the line is the box of crayons?**

Ordinal numbers on the abacus

1. Ask the children to build the stairs on their abacuses. Next ask, **How many beads are entered on the first wire?** [1] **How many beads are entered on the second wire?** [2] Continue to the tenth wire.

Ask the children to touch each wire and say the names after you: first, second, third, and so forth.

Now ask the same questions, but not in order, **How many beads are on the sixth wire?** [6] **How many beads are on the second wire?** [2]

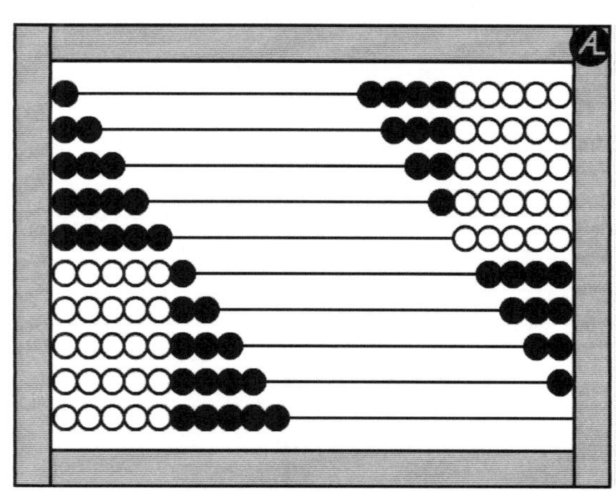

As a variation, ask the children to build the stairs backwards; that is, starting with 10. Then ask what is on the
fourth wire? [7]

2. Ask the children to, **Enter three on the seventh
wire.** Next say, **Enter six on the ninth wire; enter
5 on the first wire.** Continue until all ten wires have
quantities entered. Then ask, **How many beads are
entered on the third wire? Which wire has 6
beads?** [9th] Let the children take turns asking the other
children.

Writing ordinal numbers
Write "1st" and explain that, **This is one way to write
first. We use the number and the last two
sounds in the word, fir-*st*. How do you think
we write seco-*nd*?** [2nd] Continue with the remaining
words.

MATCHING. Copy and cut apart the ordinal number
cards and position cards from the appendix. The children
can match the two sets and play the Memory games.

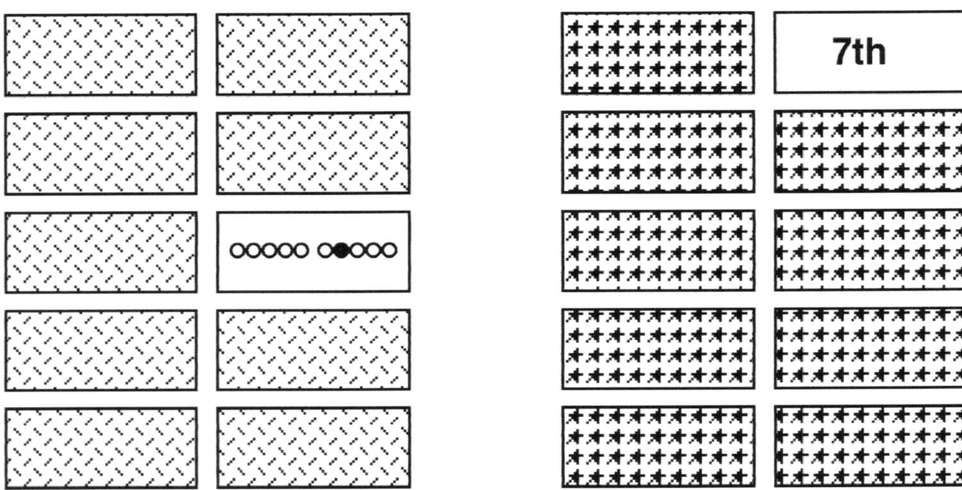

Also these games, you might to give them worksheets
(1-15 to 1-17) for matching on paper.

Applications (future review)
1. With no more than ten children in a line, ask those seated questions about unique characteristics such as, **Which
child is wearing green shoes?**

2. With the children in line, ask them to take their seats by
saying, **The fourth person may sit down.** Be sure
the beginning of the line is established before starting.
Continue until all are seated.

3. Ask questions such as, **What is the second day of
the week?** [Monday] **What is the fifth letter of the
ABC's?** [e]

Unit 2
Introducing addition

Addition at a sophisticated level is often thought as a union of two sets. However, for children it is much easier to think of it as starting with a given number and increasing it by a certain amount. Adding, when performed mentally or with a calculator or computer, requires starting with one number and then changing it. This model is also more consistent with young children's experiences; they add more food to their plates; peanut butter is added to bread; color is added to a picture.

The horizontal form is used initially to write the addition problems for several reasons. First, the work on the abacus at this point correlates directly with the horizontal form. Second, the horizontal form should be considered the norm; it is the one used in the vast majority of applications and in all mathematics beyond arithmetic. Third, children are usually learning to read at the same time and this form helps reinforce the left to right orientation. The vertical form, needed only for manual computation, is introduced later.

SIMPLE ADDITION

Before beginning to add, the children must understand some necessary terms.

Beginning adding

1. Teach the meaning of EQUAL. Use a set of hand cards and a set of abacus cards. Show the 3 from both sets and say, **What do these cards both show?** [3] **They show the same, so we call them equal. What do we call them?** [equal] Show another equal set, **Are these cards equal?** [yes] Now show, for example, an 8 from one set and a 4 from the other set. **Are these cards equal?** [no] Show other examples of equal and unequal cards. **What does equal mean?** [the same]

Here is how we write "equal": two short lines that are equal. Write the equal sign. Ask the children to write it with their fingers or on the chalkboard.

Teach the meaning of the word ADD. Place two stacks of books where the children can see. Point to one stack and say, **We are going to add more books to this stack.** Take a few books from the second stack and add them to the first stack. Call upon a child to add to it.

Repeat the activity with other objects, such as blocks in a bucket.

Teach the word PLUS with examples such as: teacher plus students equals class or players plus coach equals a team. Or try, left shoe plus right shoe equals a pair of shoes. Another is pages plus covers equal a book.

Here is how we write "plus." One line goes down and the other line goes across. Write the plus sign. Ask the children to write it with their fingers or on the chalkboard.

2. Explain to the children that you are going to add on the abacus. **I want to add 3 plus 2. First I will enter 3 on the abacus. Next I will place my left finger.** Place your left index finger after the 3. **Now I will add 2; let's count, 1, 2. Let's add them together.** Remove your finger and slide the beads together. **How much is it now?** [5] Summarize by saying, **So 3 plus 2 equals 5.**

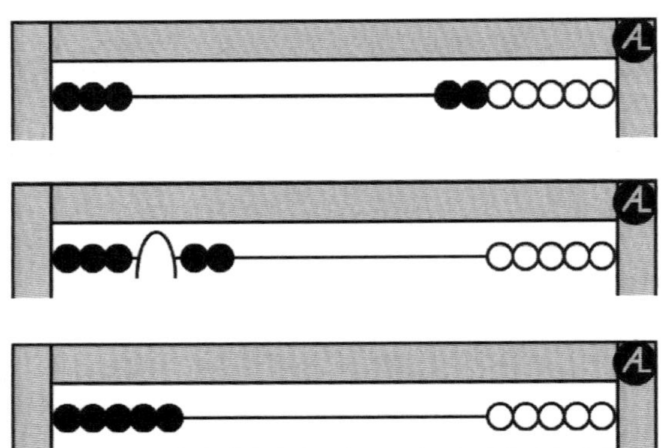

Guide the children through the same addition: enter 3, place left finger, add the second number, remove your finger, and add.

Give the children several more combinations to add, such as 5 + 3, 4 + 1, 5 + 5, and 4 + 3; be sure the sums are no greater than 10. Initially, to avoid errors, the second number should be not greater than the first number.

3. Now show the children how to write these statements on paper. **Let's write five plus two.** Use the horizontal form with rectangles on paper, an overhead projector, or a chalkboard while saying, **Five plus two; the plus sign looks like this.** Write 5 + 2. **Let's add it together and find out much it is.** Add it on the abacus. **Now we can write equals 7.** Write = 7.

Give the children another example. **Ask a child to write seven plus three.** Ask all of them to add it and another child to write the "=" sign and the sum. Repeat for 8 + 1.

Write 4 + 3 = and ask the children to add it. Call on a child to write the sum. Continue for 8 + 1 and 6 + 4.

Practice

Give the children a page of problems (2-1 to 2-3) to do. Some problems can include a number plus 0 or 0 plus a number. For a child confused by a whole page of problems, cut them apart and staple them into a booklet.

Most children need to do a page of these problems for a week or so before going on to other adding. Using the finger to separate the quantities is only a temporary measure; let the children omit it when they are ready.

By concretely performing the addition, children will better understand the operation, in contrast to workbooks where the children merely count the objects and write numbers.

ORAL PROBLEM. Part/whole circles greatly help children setup problems, ready to solve on the abacus. For each number in the problem, ask whether it is a part or a whole. Ask the children to write it in the appropriate circle. Then they find the missing amount and write it in the remaining circle. A large set is in the appendix.

Jeremy has 4 books; Jarvis gives him 3 more. How many does he have now? [7 books] One wire can represent Jeremy's books; the 3 that Jarvis gives him is added to his 4 books.

3	+	2	=	
4	+	3	=	
5	+	1	=	
3	+	1	=	
8	+	2	=	
4	+	1	=	
5	+	2	=	
1	+	3	=	
3	+	7	=	

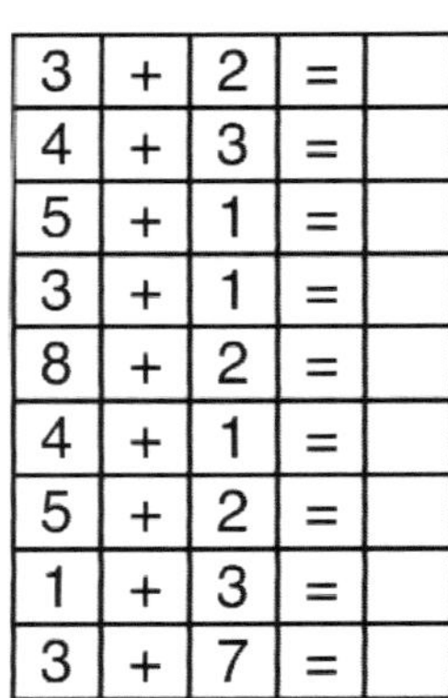

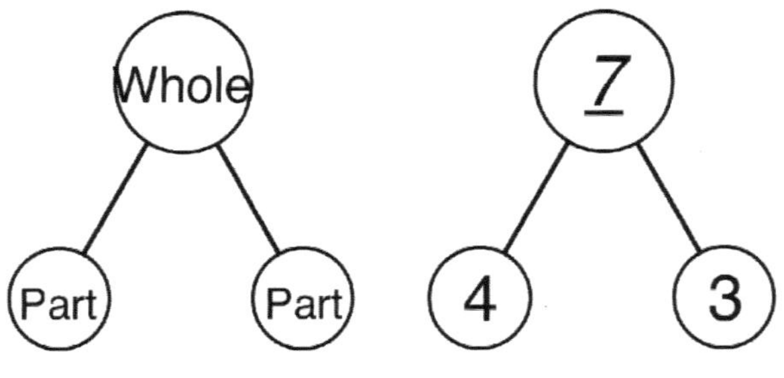

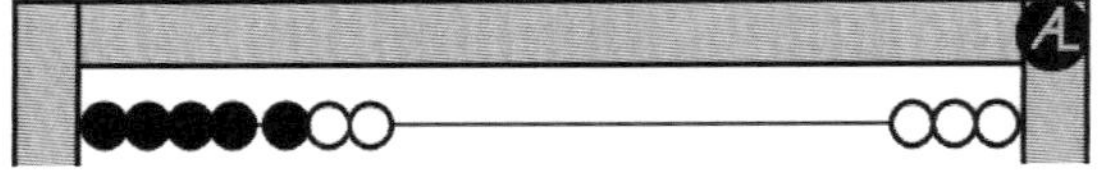

ADDING 1

Although it is obvious to us that adding 1 gives the next higher number, children need to learn it by discovery.

1. After the children have become familiar with simple addition, give them problems (2-4 and 2-5) where 1 is added to a quantity, for example, $3 + 1$, $8 + 1$, and so on.

To encourage them in discovering the rule, ask each child individually after they finish the pages if they know the "trick" or "secret or pattern" for finding the answer. Some children take a long time; whereas others are much quicker. It is vitally important that children think about what they are doing and not be satisfied with memorizing a rule.

Most children, once they know the rule, will prefer to work their problems without the abacus. They should do at least one page after discovering the rule.

2. Ask the children to build the stairs. Give them blank worksheets (2-6) with nine rows. They are to add 1 to the quantity on each wire (except the last wire) and record the result.

3. ORAL PROBLEM. Sarah has 8 dolls and receives another one as a gift. How many does she have now? [9 dolls]

One of the steps in problem solving is understanding the problem with either concrete materials or a drawing. Ask the children to show the problem with their abacuses. For this problem, they enter 8 and 1 more. Finally they write their work on paper.

THE COMMUTATIVE LAW

Another concept perfect for self-discovery is the commutative law, which for addition means that two numbers can be added in either order.

As with learning any new term, either the concept or the word must be familiar before they can be easily linked together. In other words, attempting to teach the term *commutative* before the concept is thoroughly understood results in frustration and poor learning. Therefore, do not mention the word at this point.

Reversing the addends

1. Have the children work with partners, one designated as "B" and the other as "C." Write "4 + 5 =" and say, **The Bs add four plus five on the top wire.** Next write "5 + 4 =" and say, **The Cs add five plus four on the second wire.** By writing and saying the statements, the children are learning both aurally and visually while the abacus provides the tactile stimulation. **What do you notice?** [They are equal.]

5	+	1	=	
4	+	1	=	
6	+	1	=	
1	+	1	=	
8	+	1	=	
7	+	1	=	
2	+	1	=	
9	+	1	=	
3	+	1	=	

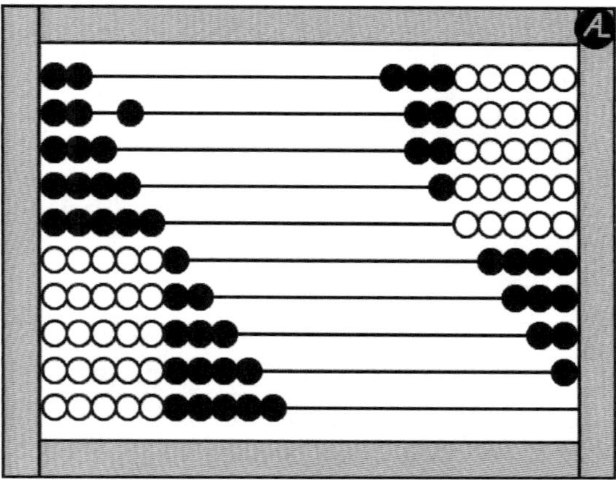

1	+	1	=	2
2	+	1	=	

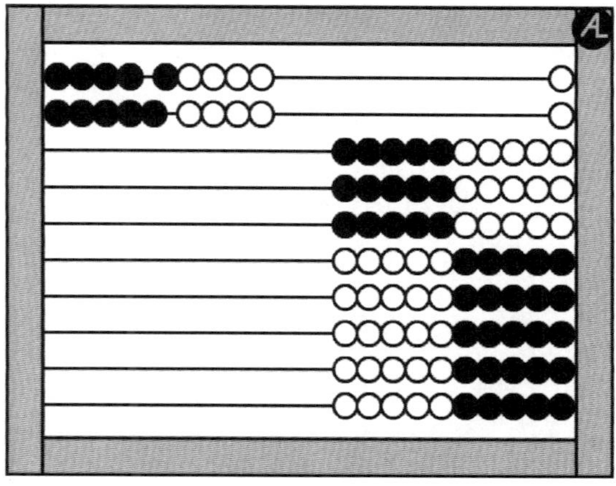

Repeat with the Cs adding 4 + 6 and the Bs adding 6 + 4. **Do you think this is always true?** Ask the children for two numbers to try.

2. Prepare worksheets (2-7 to 2-9) with problems such as 4 + 5 and 5 + 4 in pairs and let the children work independently. Observe whether or not they are using the abacus for the second sum. It takes time for children to prove to themselves that it is always true.

3. For children who have mastered adding 1 to a number and the commutative law, a worksheet with 1 plus a number will be a challenge. Let each child decide whether or not to use the abacus.

4	+	5	=	
5	+	4	=	
8	+	2	=	
2	+	8	=	
5	+	1	=	
1	+	5	=	
3	+	6	=	
6	+	3	=	

1	+	4	=	
1	+	6	=	
1	+	1	=	
1	+	3	=	
1	+	2	=	
1	+	7	=	
1	+	9	=	
1	+	5	=	
1	+	8	=	

Application

1. Adding water to containers provides an excellent example of the commutative law. Prepare four identical glasses, two filled to the same lower level and two filled to the same higher level as shown.

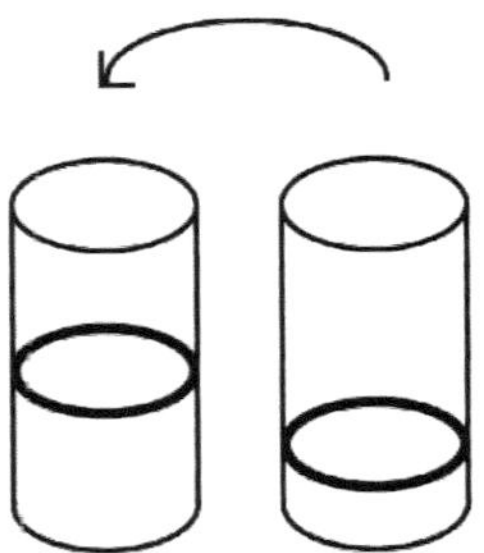 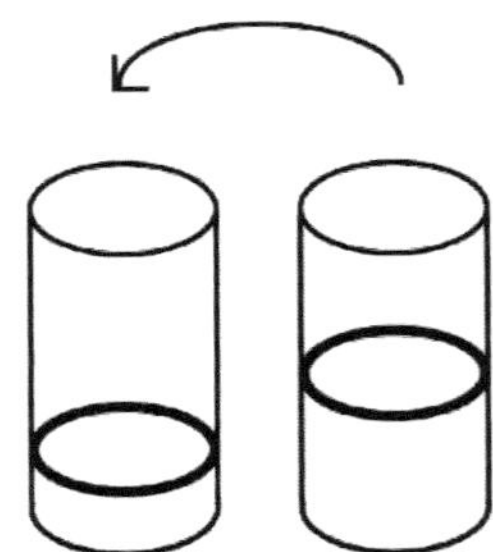

While pointing to the first arrangement, ask the children, **When I add the water from this container, how high will the water be?** Have the children venture a guess before pouring the water.

Then ask the same question while pointing to the second arrangement. [The levels will be the same.] Set up the experiment at a learning center where the children can perform it for themselves.

2. Addition problems can be worked with water using a small marked beaker, a larger marked beaker, and a pitcher of water.

THE WAYS TO MAKE 10

Ten is an important number in our number system and the facts that make 10 are also very helpful to know. Several games will be presented to help the children master these facts in a non-stressful atmosphere.

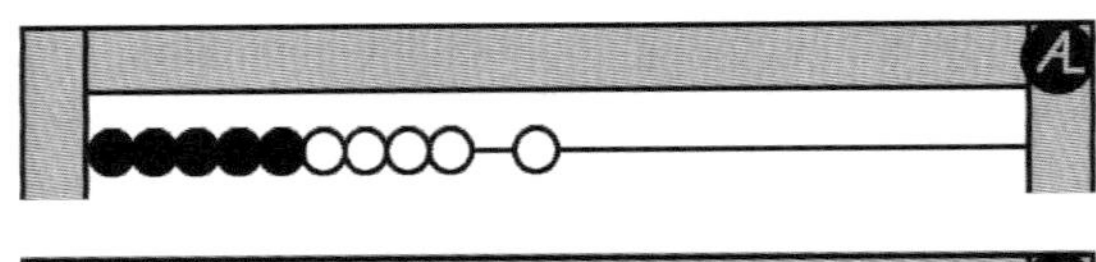

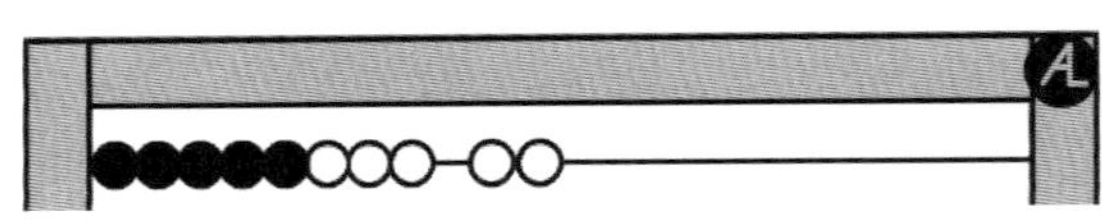

Splitting 10

1. Enter 10 on the top wire. Separate the right bead by one-half inch (1 cm) and ask a child to state the fact. [9 + 1 = 10] Move over another bead and ask another child to state that fact. [8 + 2 = 10] Continue to 1 + 9.

2. Give the children a blank worksheet and ask them to use their abacuses and record the combinations that make 10. After they are written, have the children find the "twins"; for example, 1 + 9 and 9 + 1. Each set of twins can be circled with a different color. Or the facts can be cut apart and the twins pasted together on another sheet of paper. One fact, 5 + 5, is called a "double" and does not have a twin.

	+		=	
	+		=	
	+		=	
	+		=	
	+		=	
	+		=	
	+		=	
	+		=	
	+		=	
	+		=	

9	+	1	=	10
8	+	2	=	10
7	+	3	=	10
6	+	4	=	10
5	+	5	=	10
4	+	6	=	10
3	+	7	=	10
2	+	8	=	10
1	+	9	=	10

Games

1. HANDSHAKING GAME. Distribute among ten children one of each number 1 to 9, but two 5s. Build the stairs on the abacus; move the remaining beads near the stairs to narrow the gap, as shown. Touch the top wire and read the fact, **1 + 9**. The children having those numbers shake hands gently. Continue to 9 + 1. Then ask, **Who shook hands twice?** [All but those with the 5s] **Who shook hands only once? Why?**

2. FINDING THE PAIRS. This game for two or three players uses the same cards as the Handshaking Game. Spread the cards out face up. Without regard to turn, each player picks up a card, for example 2, and enters that number on the abacus. Next he slides over the remaining beads in the row leaving a finger's width. Ask, **How many beads are needed to make 10?** [8] Tell the child to find the 8-card and to set the pair aside.

3. MATCHING MEMORY. Use the same cards to play Memory. When the first card is turned over, the player states what card is needed to make 10 before turning over the second card.

4. GO TO THE DUMP. The pairs for this game, played similar to Go Fish, are the combinations that make 10.

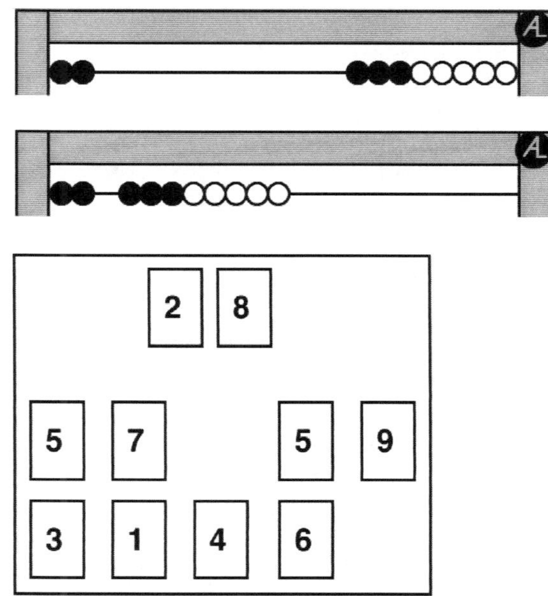

Missing addends

1. Write 7 + __ = 10 and say, **Seven plus what equals ten.** Put 7 on the top wire and 10 on the second wire. **How many beads do you have to add to make these equal?** [3] The children can either count the missing spaces or count the beads as they move them over. **So seven plus three equals ten.** Write the 3 in the blank. Repeat for another example: 4 + __ = 10.

Give the children a prepared worksheet (2-10 and 2-11).

2. ORAL PROBLEMS.

A. Carlos is trying to get 10 gold stars. If he already has 6, how many more must he receive? [4 stars]

B. Kay has 10 houses on her block. If she has walked by 6 houses, how many more must she walk by before she reaches the end of the block? [4 houses]

C. Tracy learned the names of 2 more of her classmates. She already knew 5 names. Now how many names does she know? [7 names]

D. What even number comes after 6? [8]

E. What odd number comes after 7? [9]

F. What even number comes before 6? [4]

E. What odd number comes before 7? [5]

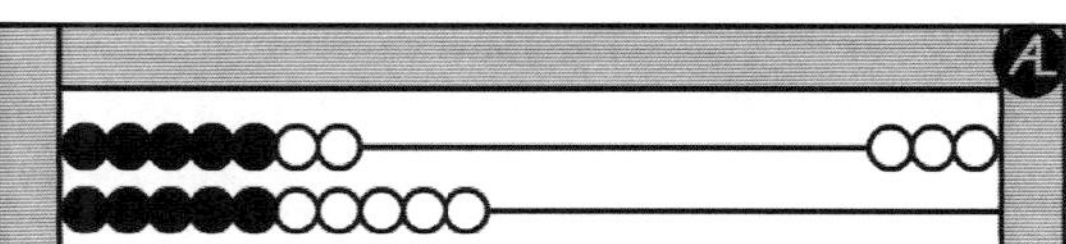

7	+		=	10
4	+		=	10
9	+		=	10
2	+		=	10
5	+		=	10
8	+		=	10
1	+		=	10
6	+		=	10
3	+		=	10

SPLITTING NUMBERS 9-2

For the sake of completeness, the children can split the other numbers and work problems with missing addends. Strategies to memorize these facts will be presented later.

More splitting

Ask the children to enter 9 on their abacuses. Then ask them to separate 1 bead a finger's width and to recite the fact. [8 + 1 = 9] Separate another bead and recite that fact. [7 + 2 = 9] Continue to 1 + 8 = 9. These facts can be written on blank worksheets. Then ask the children to circle the twins. Ask them if 9 has a double. [no]

Similarly, the children can write the ways to make 8, 7, 6, 5, 4, 3, and 2 and circle the twins. Ask them which numbers have doubles. [the evens] When all the facts are written, the separate sheets can be assembled to form a booklet.

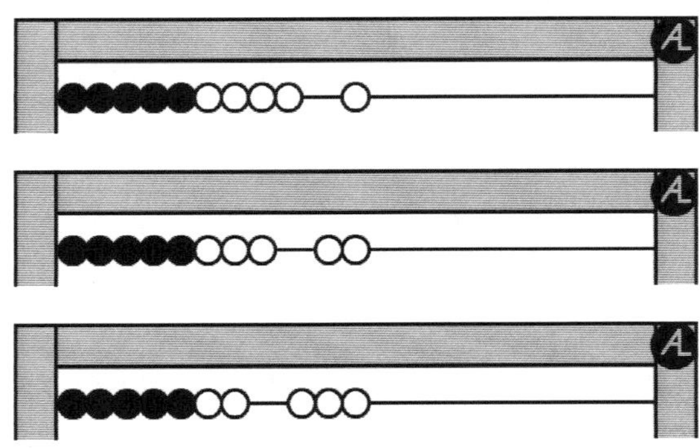

Missing addends

Write and say, **5 + __ = 7 (Five plus what equals 7.)** Enter 5 on the top wire and 7 on the bottom wire. **How many beads do we need to make the first wire equal to the second wire?** [2] Write the 2 and say, **Five plus two equals seven.** Repeat for 3 + __ = 8. Encourage the children to take a guess and then check their guess. Guessing and checking is an important skill that needs fostering.

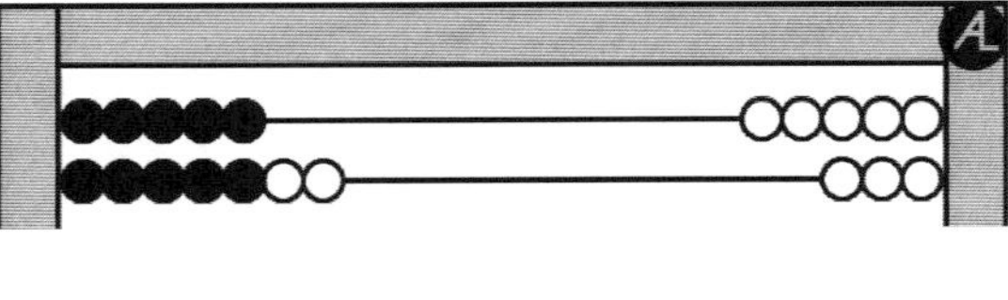

Ask the children to do the following examples on their abacuses before giving them a worksheet (2-12 and 2-13): 2 + __ = 3; 4 + __ = 7; and 1 + __ = 6.

ORAL PROBLEMS. A. Sarah is waiting 7 days for her family to go on a vacation. Five days had gone by. How many more days must she wait? [2 days]

B. Kevin was expecting 9 cub scouts to come to his house. If 3 had already arrived, how many more were still expected to come? [6 cub scouts]

C. Mindy's group has 9 children. If they work with partners, will everyone have a partner? [no]

D. The letters A, B, C, D, E, F, and G are lined up in a row. Which letter is in the middle? [D]

E. Rick received 4 more stamps today. He had 5 before. How many does he have now? [9 stamps]

5	+		=	7
3	+		=	10
5	+		=	6
3	+		=	5
8	+		=	9
4	+		=	6
6	+		=	8
2	+		=	3
3	+		=	9

Unit 3
Numbers from 11 to 100

This unit can be begun before the completion of Unit 2. Rather than ask the children to simply memorize the words for the numbers 11 to 31, as is often done, teach the concept of tens and ones. One objective of this unit is that the children discover that the tens are distinct from the ones, but that many of the rules governing them are the same as those for ones.

Children need to learn from the beginning in their study of mathematics that there is order and that memory will not substitute for thinking and understanding. The names forty, sixty, seventy, eighty, and ninety follow an order and should be taught before the exceptions of thirty, fifty, twenty and especially the teens.

In this unit the place-value cards are introduced. For example, the number 36 is represented by a "30" card and a "6" card. The 0 is not eliminated, rather the 6 is placed on top of the 0 of the 30. These cards virtually eliminate the problem of reversals. The children need to be encouraged to use them to form their answers before copying them to paper. A set of them may be found in the appendix; they are also available in plastic from Activities for Learning.

THE TENS

Initially, the generic term "two ten" is used rather than "twenty." This helps the children focus on the nature of the tens without being burdened with remembering the names. Use the term "2-ten" and not "2 tens" because it is easier to say and is similar to saying 3 hundred and not 3 hundreds.

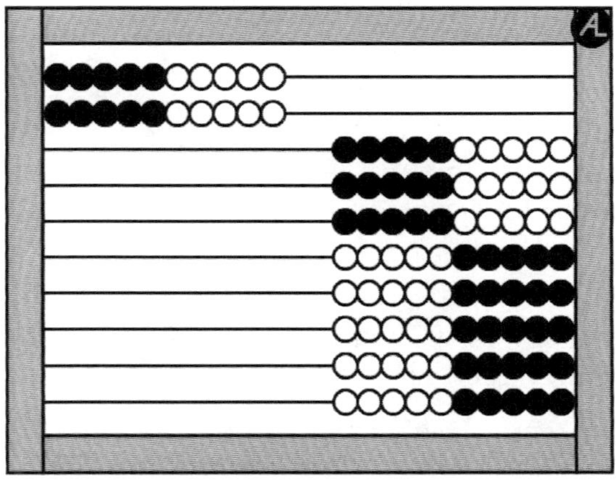

Entering tens
1. Ask the group of children, **Can you enter 1 ten?** Note that the term ONE TEN is new; previously we have said "ten" or "a ten." Clear your abacus and enter 2 tens while saying, **This is 2-ten; can you enter 2-ten.** Repeat for 3-ten, 4-ten, and 5-ten.

Enter a number of tens and ask a child to name the quantity. For example, enter 3-ten and ask, **How much is this?** [3-ten] The children can practice entering and naming quantities with partners.

2. Teach the quantities 6-ten to 10-ten in the same way. Be sure the children notice the color change after 5-ten. Ask the children to enter the quantities and to read the entered quantities.

3. Introduce the term 5 ONES. Explain that each bead is called a one, so 5 ones is 5 beads. Ask the children to **Enter 5-ten**. Then tell them, **Clear and enter 5 ones.** Ask for other quantities, such as 8-ten a 4, and 7 ones. Give each child a turn to ask the group to enter a quantity. This provides the child with opportunities to lead and to make decisions.

Reading tens

Cut apart the place-value cards 10, 20, ... 100 to be found in the appendix or use plastic cards.

1. Ask the children to enter 3-ten on their abacuses; show them the place-value card of 30. Say, **This is how we write 3-ten;** point to the 3 while saying "3" and to the 0 while saying "ten." Hold up other place-value cards and ask the children to read them. Point to the appropriate number as they read them.

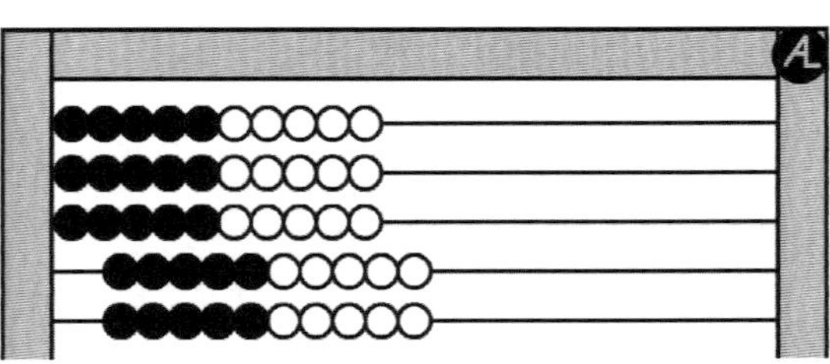

2. Hold up various place-value cards one at a time and ask the children to clear and enter that quantity on their abacuses; include cards with quantities less than 10.

3. Give each child a set of place-value cards and tell them, **Spread out your cards so you can see them.** Enter 5-ten on your abacus and ask the children, **Can you find the card that shows this quantity?** Continue with other quantities.

Give each child a turn to enter a quantity while the others find the cards. The children can also work with partners for more practice in reading and finding cards.

Adding tens

Adding tens provides the children with an opportunity to work with tens and to discover the similarity with adding ones.

1. Write "30 + 20 = " and say, **Can you add 3-ten plus 2-ten?** Continue with 50 + 30 and 80 + 20. Give them worksheets (3-1 and 3-2) similar to the one shown on the left.

2. Later, give them worksheets (3-3) where they add 10 to a group of tens. It is also helpful to give a worksheet (3-4) such as the one on the right.

ORAL PROBLEMS. Ask the children to solve similar problems: If Vanessa has 40 cents and earns 30 cents more, how much will she have now? [70 cents]

30	+	20	=	
20	+	40	=	
40	+	30	=	
50	+	10	=	
10	+	10	=	
20	+	60	=	
10	+	20	=	
60	+	20	=	
20	+	30	=	

2	+	5	=	
20	+	50	=	
1	+	6	=	
10	+	60	=	
4	+	5	=	
40	+	50	=	
7	+	1	=	
70	+	10	=	

Splitting 100

To help the children gain more facility in thinking in terms of and in writing tens, have them find all the ways to make ONE HUNDRED. Show the children the place-value card 100 and ask them to enter it on their abacuses. Explain that, **10-ten is also called one hundred; what is another name for 10-ten?** [one hundred]

Use a pencil to show the separations or move the tens about 2 cm (1 inch) away from the left edge. Start by separating 1 ten and ask a child to recite the fact. [9-ten + 1 ten = 1hundred] Separate another ten and ask another child to give the next fact. [8-ten + 2-ten = 1hundred]

The children can write the facts totally 100 on a blank worksheet (2-6).

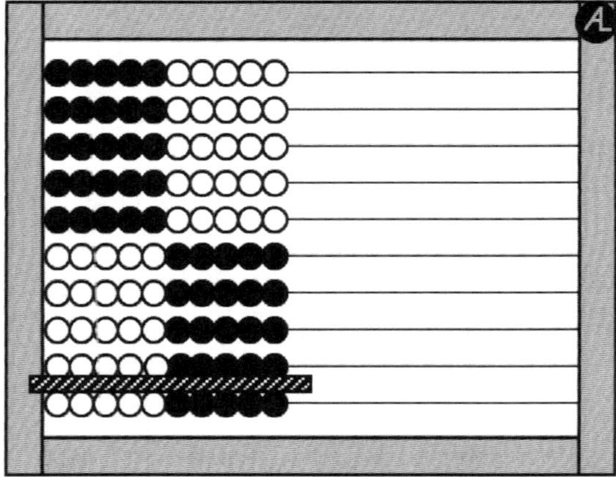

TENS AND ONES

In teaching concepts, it is important to teach first the general rule and later to teach the exceptions. Therefore, delay teaching the teens because they are read backwards; the ones are said before, instead of after, the tens.

Tens and ones on the abacus

1. Ask the children to enter two quantities: **Enter 2-ten and 6 ones.** The ones are entered on the wire immediately below the last ten. Next ask them to, **Clear and enter 4-ten and 8 ones.**

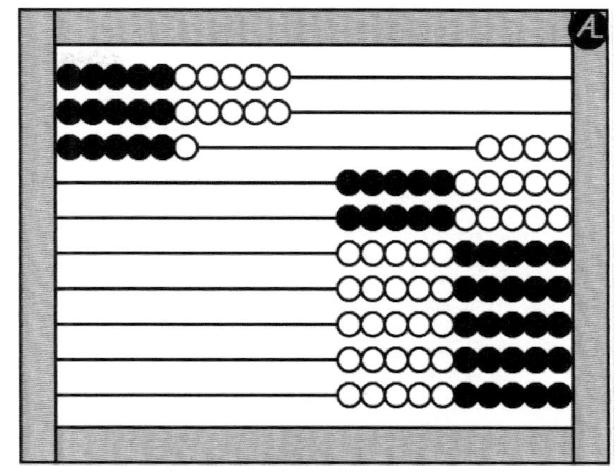

Finally, tell them to, **Enter 3-ten and 6 ones.** Point to the place-value cards of 30 and 6 while saying, **You know how to write 3-ten and you know how to write 6 ones.** Slide the "6" down and cover the "0" of the 30, **This is how to write 3-ten and 6 ones.** Note that with the place-value cards, reversals are virtually eliminated.

2. Ask the children to sit with their partners. Give each pair a set of place-value cards from 1 to 90 to be spread out. Let the children decide in each partnership who will be the "ten" person and who will be the "ones" person.

The "ten" person enters the tens quantity on the abacus and locates the corresponding ten-card. The "ones" person enters the ones, locates the ones card, and places it over the "0" of the tens card.

Give the children the following combinations to enter: **8-ten and 3 ones; 7-ten and 9; 1 ten and 6; and 5 tens and 2.**

Ask the children to change roles. Then give them these combinations: **9-ten and 4; 2-ten and 7; 6-ten and 1 one; and 4-ten and 5.**

Practice

1. TENS AND ONES BINGO. This game can be played with the entire group or with smaller groups of three to four. Each player, with the exception of the one caller per game, chooses and lays out 16 place-value cards from 1 to 100 in a four by four array.

The object of the game is to overturn all the cards in either a row, column, or DIAGONAL. Informally explain that a diagonal is a line from one corner to the opposite corner. The caller shuffles the cards and reads each by saying, for example, 4-ten. A player having that card turns it face down. A player signals he has won by saying the appropriate term, "row, column, or diagonal." Later, the caller can give combinations like: **9-ten and 4 ones; 2-ten and 7; and 6-ten and 1.**

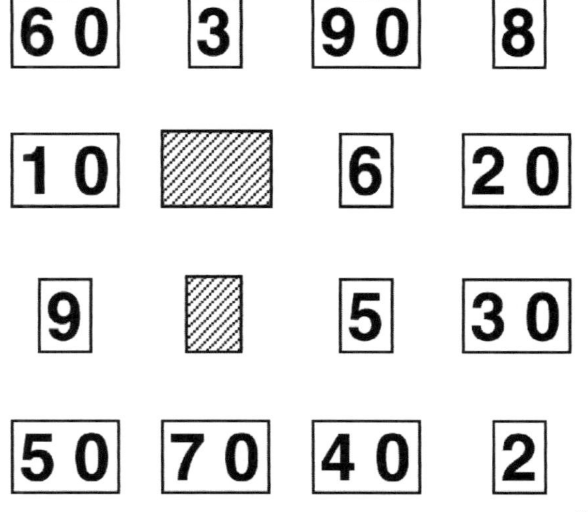

Adding tens and ones

Although it may seem trivial to us to do the worksheet (3-5) with sums like 30 + 4 and 70 + 8, most children find it a challenge; it helps them to discover the relationship between tens and ones. For real understanding, it is vitally important that they use the place-value cards initially for this work.

30	+	4	=	
70	+	8	=	
60	+	2	=	
40	+	6	=	
50	+	3	=	
10	+	5	=	
90	+	1	=	
80	+	9	=	
20	+	7	=	

MORE TENS AND ONES

At this point the usual names are introduced, as is counting by tens and adding in the upper decades.

Customary names

1. The children can now begin learning the customary names. Start with forty: **4-ten has another name; it is also called forty**. Repeat for 6-ten. Let them guess the names for 70, 80, and 90. Then teach thirty and fifty referring to the ordinal numbers third and fifth. Lastly, teach twenty, which can be thought of as "twin ten."

2. CAN YOU FIND. Each player spreads out the place-value cards from 1 to 90. Explain that you are going to name a number and they must find the cards needed to build that number as fast as possible. The cards are then placed along the edge of the table or rug. Proceed through the four levels of difficulty as the children are ready.

a) Ask for one card by generic name; for example, 4-ten, 6, 2 ones.

b) Ask for one card by customary name; for example, eighty, twenty, two.

c) Ask for two cards by generic name; for example, 7-ten and 1, 4-ten and 3.

d) Ask for two cards by customary name; for example, thirty-six, sixty-eight.

Change the degree of competition by the tone of your voice.

Counting by tens

Counting by tens on the abacus further helps the children correlate the names as well as memorize the sequence. First enter a ten and say, **Ten.** Enter another ten and say, **Twenty.** Enter another ten and say, **Thirty.** Continue to one hundred. When they are ready, ask them to do it without the abacus.

Tens and ones + ones

1. Write "46" and ask the children to build it with their place-value cards and then enter it on their abacuses. Call upon a child to read it. If the child experiences difficulty reading it, hold up place-value cards; remove the 6 from the 40; wait until the child has read "forty"; and then replace the 6. Repeat with other numbers such as 78, 92, 36, and 27.

2. Next write "43 + 2." Guide the children through building and entering the 43 and adding the 2. The sum should be shown with the place-value cards before it is written. Prepare worksheets (3-6 and 3-7) with similar problems.

3. Also prepare worksheets (3-8 and 3-9) with combinations like 34 + 6 and 69 + 1. Guide the children to discover the rule that if the ones equal 10, the sum is the next higher ten.

4. It is also helpful to give the children a page (3-10) where the units remain the same; for example, 23 + 6, 63 + 6 and the like.

ORAL PROBLEMS. A. In Kari's school, there were 63 first graders. Five more moved in. How many are there now? [68 first graders]

B. Miss Green's class knew 24 songs from last year. This year they learned 6 more. How many do they know now? [30 songs]

C. Lucas knew 27 states by shape. Today he learned 3 more. How many does he know now? [30 states]

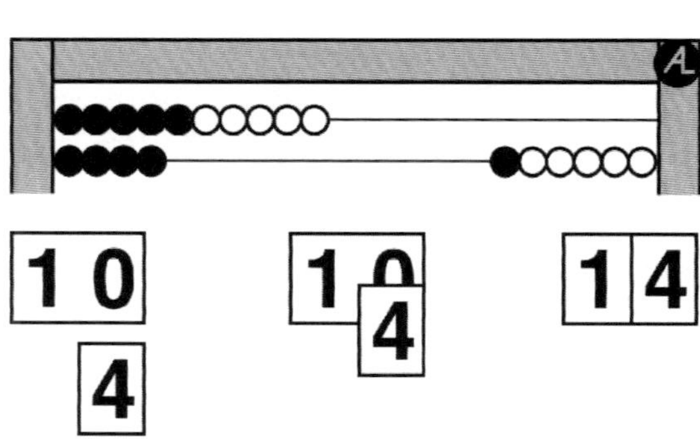

43	+	2	=
21	+	6	=
65	+	3	=
84	+	1	=
52	+	4	=
76	+	3	=
94	+	5	=
85	+	1	=
53	+	3	=

34	+	6	=
69	+	1	=
56	+	4	=
21	+	9	=
47	+	3	=
82	+	8	=
75	+	5	=
18	+	2	=
22	+	8	=

23	+	6	=
63	+	6	=
83	+	6	=
73	+	6	=
13	+	6	=
33	+	6	=
93	+	6	=
53	+	6	=
43	+	6	=

THE TEENS

Strictly speaking, the term teen includes only numbers 13 to 19, but the activities here are intended for numbers 11 to 19. The names of these numbers 11 to19 do not follow the general rule and therefore need special work.

The traditional names

Since the teen numbers are pronounced in reverse order, that is, ten 4 becomes fourteen, the following word game is helpful. Say a word like bedroom and ask the children to say it backwards. [room-bed] Repeat for fireplace [place-fire and paper-news. [newspaper]

14: Ask the children to enter ten-4 on their abacuses. **This number has another name, fourteen.** Ask the children to point to the four while saying "four" and to the 10 while saying "teen." **Does it sound backwards?** [yes] Also tell them, **The word teen means "ten."**

16-19: These are similar to 14.

13, 15: Teach 13 like 14, but with the reminder that we have to say "thir" as in third grade or thirty. Repeat for 15, which has the "fif" sound.

11: Tell the children this story. About 1000 years ago, some people looked at ten 1 and said they thought a good name for the number would be "a one left" because it was 1 left over 10. However, they said it backwards, "a left one." Soon that became "eleven." This is the derivation of the word

12: Enter ten 2 and tell the children that these same people decided to call it "two left," which became "twelve."

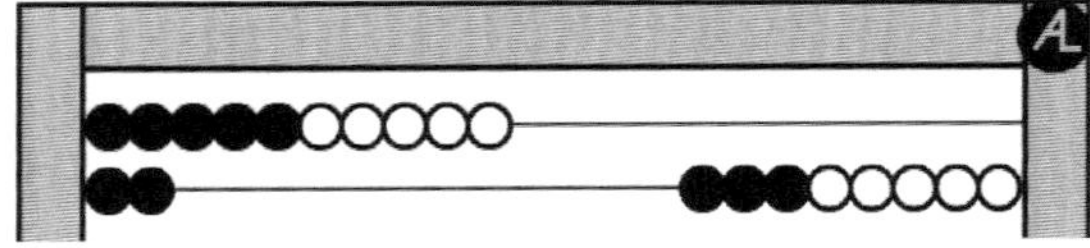

Practice

1. Enter one of the teen quantities on your abacus. Ask the children to build it with the place-value cards and name it. Give each child a turn to enter one of the quantities for the other children to build and name.

2. Have the children work with partners to practice. One child enters a number and the other child names it. Then give them cards for playing Matching Memory. They can also match on worksheets (10-11 and 10-12).

3. Say a quantity and ask the children to enter it and build it. Let each child name a quantity.

4. Write one of the numbers and ask the children to enter it and read it. Let each child write a number.

5. For readers, prepare cards with the names written out. The children can match them with corresponding number cards and play the Memory games. Then give them worksheets (10-13) for matching.

6. Have the children count to 20 starting with one bead and saying, "One." Continue to enter one bead at a time and name the quantity.

7. Play the Can You Find game. Remind the children when giving a number, for example, 14, that teen means ten. Otherwise, when they hear the four in fourteen, they are apt to look for the 40 card.

8. On a large clock, have the children read the numbers in order.

9. On the clock, cover up a number and ask the children to name it.

10. Play the clock games found in *Math Card Games*.

Adding with sums in the teens

1. The first adding is similar to that performed with the tens and ones. Use the worksheet (3-14) with 10 + numbers 1 to 9 as shown on the right.

10	+	3	=	
10	+	8	=	
10	+	6	=	
10	+	7	=	
10	+	9	=	
10	+	4	=	
10	+	2	=	
10	+	5	=	
10	+	1	=	

Sometimes, children who learned to count by rote before understanding tens and ones will want to count all the beads to determine the sum in the teens. This helps them incorporate their new knowledge with the old. Discourage it after a couple of times.

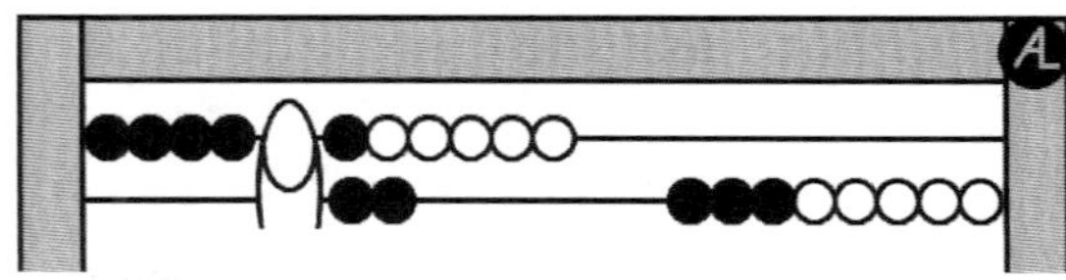

2. Write 4 + 8 on the board and tell the children, **This is how we add 4 + 8**. Start by entering 4 on the first row of the abacus and placing a finger after it. Add the 8 by first completing the top row with 6 and then entering the remaining 2 on the second row, all to the right of the finger. Finally, the sums are added together by moving the beads to the left.

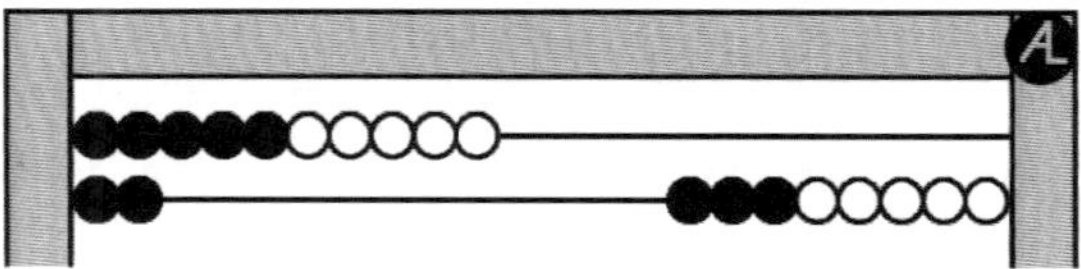

Ask the children to add other examples, such as 6 + 7 and 5 + 9. Check to make sure they complete the first row and not enter the entire second addend on the second wire.

Give worksheets (3-15 to 3-18) with the sums in the teens as shown.

3. ORAL PROBLEMS. A. A townhouse has 8 windows on the first floor and 7 windows on the second floor. How many windows are there in all? [15 windows]

B. What number is 6 more than 10? [16]

C. Alisha wants 9 balloons for her room and 3 for school. How many should she buy? [12 balloons]

D. Cody saw 9 cars with out-of-state licenses on the way to a picnic. On the way back he saw 8. How many did he see that day? [17 out-of state licenses]

4. Problems like 39 + 6 are an extension of sums in the teens. The 39 is entered first and then the 6 is added as above. Success in multiplying requires the ability to perform this step mentally. Give them similar worksheets (3-19 and 3-20).

It is also instructive to give pages (3-21 and 3-22) where the ones stay the same while the tens vary.

4	+	8	=	
5	+	8	=	
8	+	6	=	
9	+	5	=	
7	+	4	=	
9	+	2	=	
6	+	6	=	
4	+	9	=	
8	+	7	=	

5	+	1	=	
5	+	2	=	
5	+	3	=	
5	+	4	=	
5	+	5	=	
5	+	6	=	
5	+	7	=	
5	+	8	=	
5	+	9	=	

39	+	6	=	
83	+	9	=	
26	+	9	=	
46	+	8	=	
66	+	5	=	
37	+	7	=	
77	+	4	=	
19	+	8	=	
49	+	5	=	

86	+	9	=	
6	+	9	=	
36	+	9	=	
76	+	9	=	
46	+	9	=	
16	+	9	=	
26	+	9	=	
66	+	9	=	
56	+	9	=	

Adding double digits

The children are now capable of adding quantities such as 37 + 29, where the sum is not over 100. Write on the board

$$37 + 29 =$$

and ask the children to enter the 37 with the ones separated, that is, near the bottom. Next ask them to add the 2-ten and finally the 9 ones. If desired, they can use take and give to relocate the new ten with the other tens.

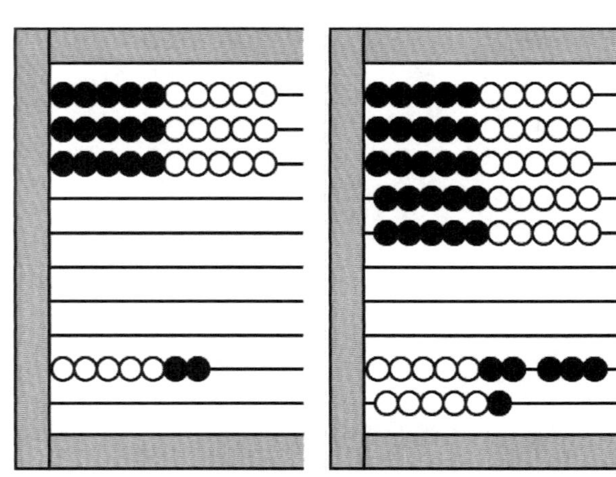

Give them examples 45 + 23 and 39 + 35 before assigning a worksheet (3-23).

ORAL PROBLEM. A. Jamie bought a notebook for 59 cents. The sales tax was 3 cents. How much did he pay for the notebook? [62 cents]

B. Dawn rode 30 miles to visit her grandmother and 30 miles back home. How many miles did she ride both ways? [60 miles]

37	+	29	=		
45	+	18	=		
23	+	48	=		
34	+	27	=		
65	+	21	=		
15	+	15	=		
36	+	47	=		
22	+	38	=		
54	+	19	=		

VERTICAL ADDITION

The even-odd patterns with the abacus turned ninety degrees provide another way of adding on the abacus. It will be used extensively later in the trading, or carrying, process. Before beginning this lesson, review the names of the quantities in the even-odd format.

The abacus in this position is a little more difficult for demonstrating since the beads tend to fall down. Therefore, when holding the abacus vertically, hold up the beads with one thumb held behind the abacus and use your other thumb to add the second quantity.

Adding with sums ten or less

1. With their abacuses "sideways," ask the children to enter 4 and say, **Now we will add 6 more.** Show them how to slide the 6 upward as a group by using either a thumb or two fingers. Momentarily maintain the space between the quantities with your thumb, then remove it and slide the beads together.

Give them other oral examples, such as 8 + 2 and 2 + 5. For now, use only evens for the top number and keep the sum equal to ten or less.

Write on the board

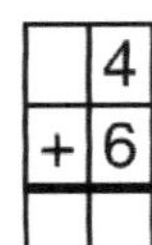

and tell the children, **This is another way to show adding. The line underneath means equal.** Point to it. **The sum is written below the line.** Write 10 with each digit in its separate box.

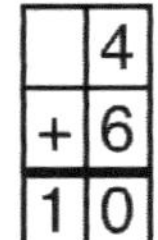

We read it like this: four plus six equals ten. Point to each number or symbol as you say it. Then change the numbers to 4 + 5 and ask the children to add it. Ask one of them to write the sum and another to read the entire problem.

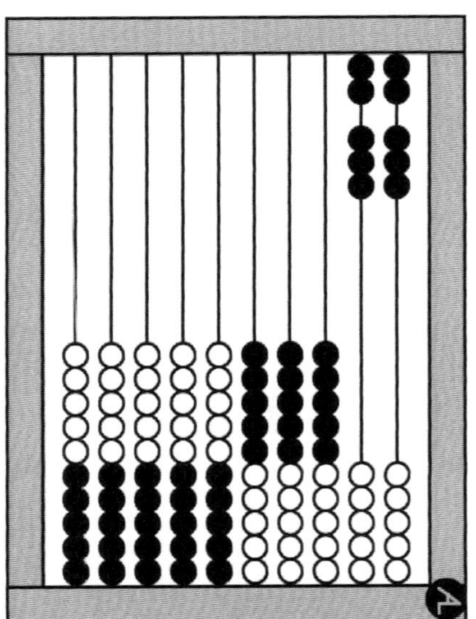

4	2	8	4
+ 6	+ 4	+ 1	+ 2

6	2	8	6
+ 1	+ 3	+ 2	+ 4

2	2	4	8
+ 1	+ 6	+ 5	+ 0

2. Give the children worksheets (3-24 and 3-25) with similar problems.

3. Write it in the vertical format and say, 3 + 5.

Explain that to keep the rows as even as possible, the last bead for the 5 is taken from the right wire. Repeat for 5 + 5 and 7 + 2.

Write 6 + 3 on the left side of the board in the horizontal format and write it on the right side in the vertical formal.

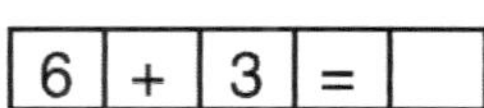 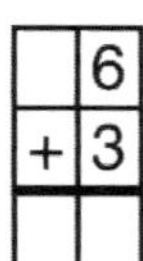

Ask the children, **Do you think the answers will be the same?** [yes] Either have half of the children add it "the old way" and the other half add the "the new way," or have all of them do it both ways.

Give the children prepared worksheets (3-26 and 3-27).

Adding with sums over ten

There is very little difference in vertical addition for sums between 11 and 18. The actual adding is identical. The objective is that the children be able to read the sum at a glance by observing that the dark beads are a ten and that the number of light-colored beads entered is the ones.

1. Write and say 10 + 3 in the vertical format. Ask the children to add it and find the sum. Make available the place-value cards for those needing them. Remind them that only one digit may be written in each box. Give them worksheets (3-28 and 3-29).

2. Write, using the vertical format, and say, **Add 7 + 8.** Ask a child to write the sum in the boxes. Prepare several worksheets (3-30 to 30-32) such as the one shown.

ORAL PROBLEMS. For these problems, let the children choose the abacus orientation.

A. Samantha's class has 10 girls and 9 boys. How many children are in the class? [19 children]

B. Lee counted 6 sailboats on one side of the bridge and 8 boats on the other side. How many did he see? [14 boats]

C. Jennie helped her mother plant tulips. Her mother dug the holes and Jennie planted the bulbs. If Jennie planted 16 bulbs, how many holes did her mother dig? [16 holes]

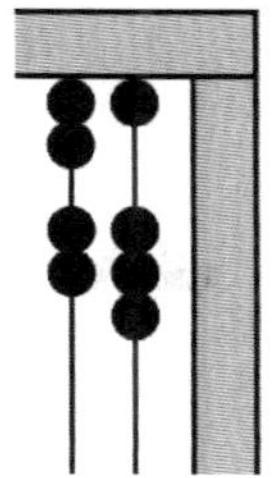
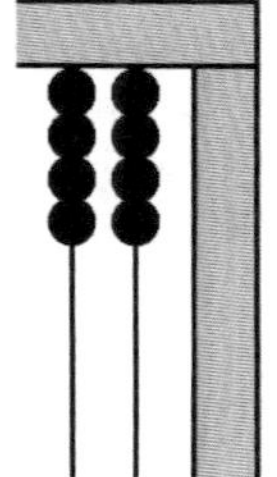
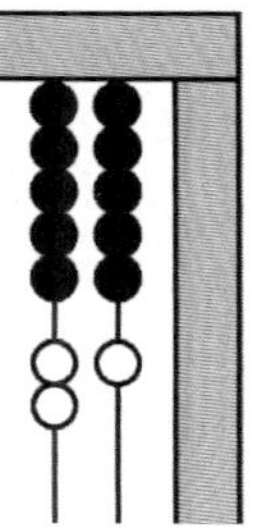
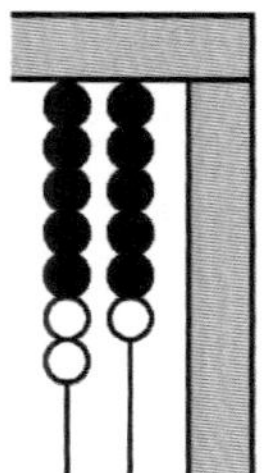
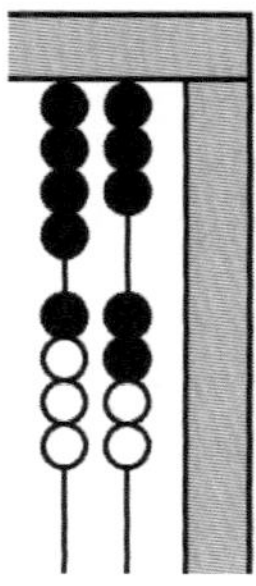
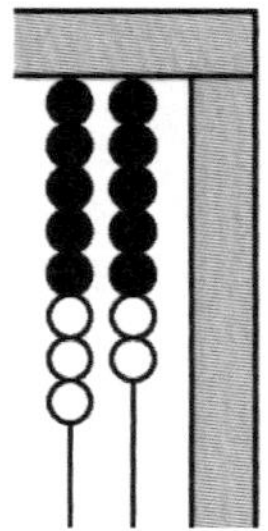

3		5		3		1
+ 5		+ 1		+ 7		+ 3

1		5		7		5
+ 8		+ 4		+ 1		+ 5

5		7		9		3
+ 3		+ 3		+ 1		+ 0

1 0	1 0	1 0	1 0
+ 3	+ 1	+ 2	+ 4

1 0	1 0	1 0	1 0
+ 8	+ 6	+ 9	+ 7

1 0	1 0	1 0	1 0
+ 5	+ 0	+ 2	+ 8

7		4		4		7
+ 8		+ 7		+ 9		+ 6

8		9		7		3
+ 4		+ 8		+ 4		+ 8

8		5		8		6
+ 6		+ 9		+ 5		+ 9

ADDING MORE THAN TWO

Frequently, three or more numbers needed to be added together.

Making Tens game
The purpose of this game is to determine the total number of beads by arranging them in groups of ten.

1. Write on the board in a column the numbers as shown. Ask the children to enter those quantities as they are read. Then say, **We want to find the sum, but first take a guess.** Write down each child's guess. Then say, **We can do this by counting no further than ten.** Ask them to watch while you demonstrate the method.

Notice that there are 4 beads on the first wire; continue counting on the second wire to 7. Continue on the third wire where 10 is reached after 3 beads. Place a finger after the third bead and finish counting the row, starting from 1. **These first three rows are equal to 1 ten and 3. So we will remove these beads and replace them with a 1 ten and 3 beads.** Clear the first three rows and enter 10 on the first wire and 3 on the third wire as shown.

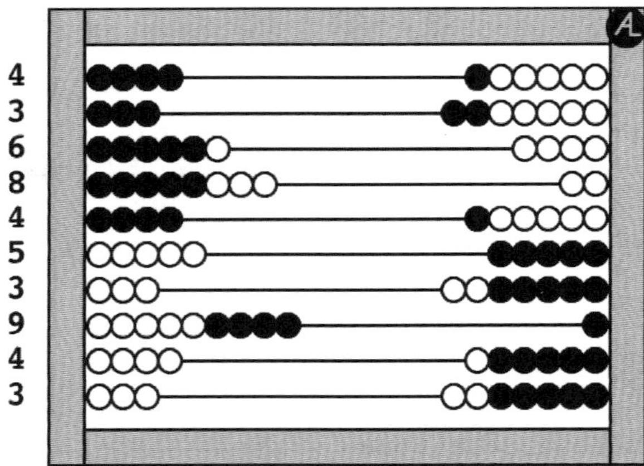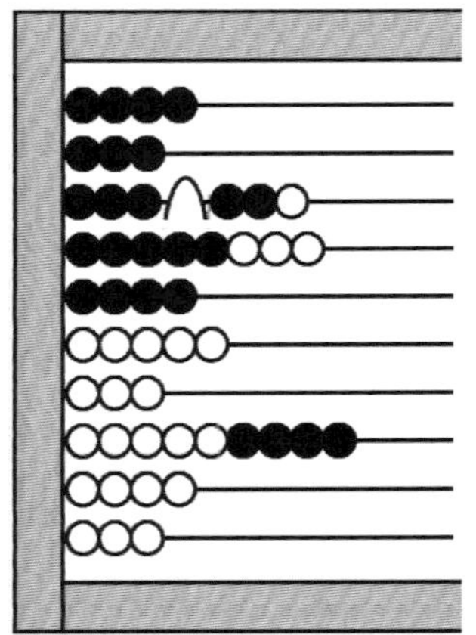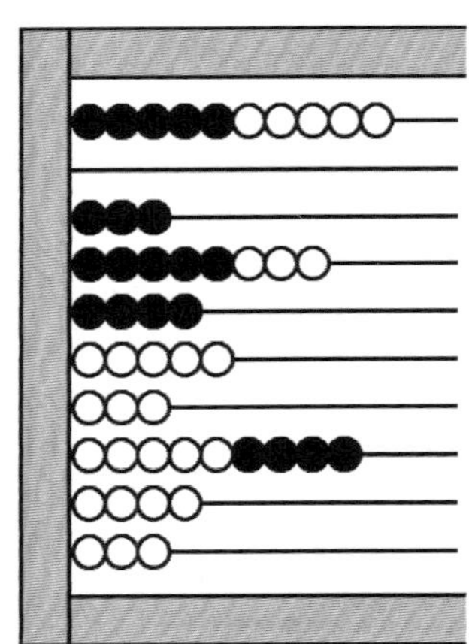

This time, start counting with the 3 on the third row and continue to the fourth row. Place a finger after reaching 10. Point to the fourth and fifth rows: **These two rows equal what?** [1 ten and 1 one] Clear the rows and

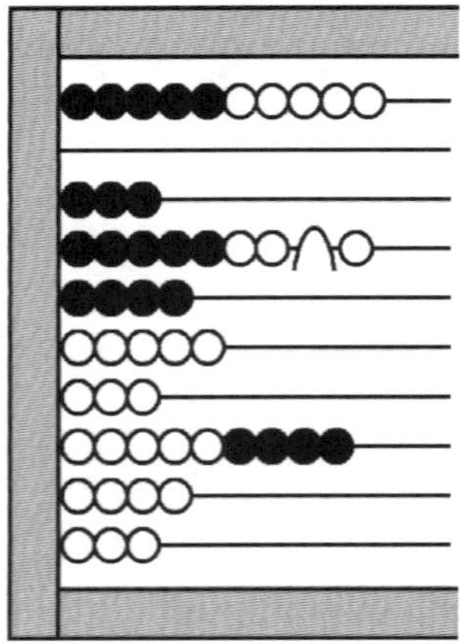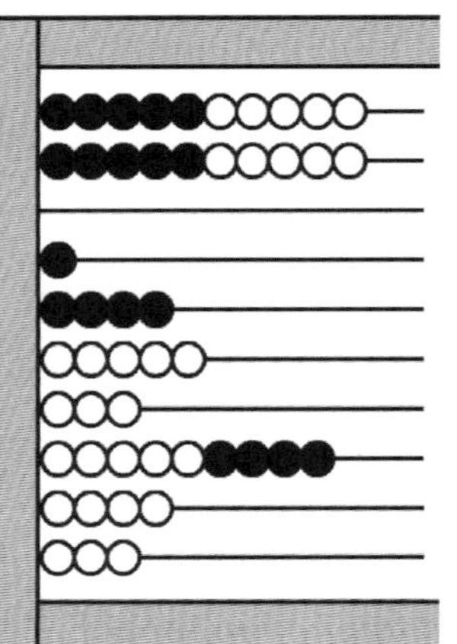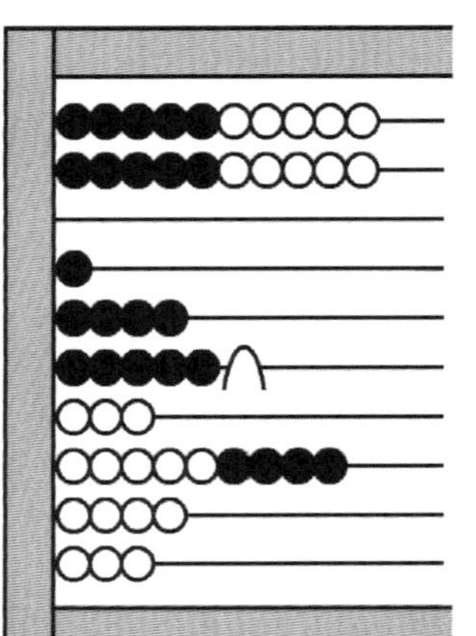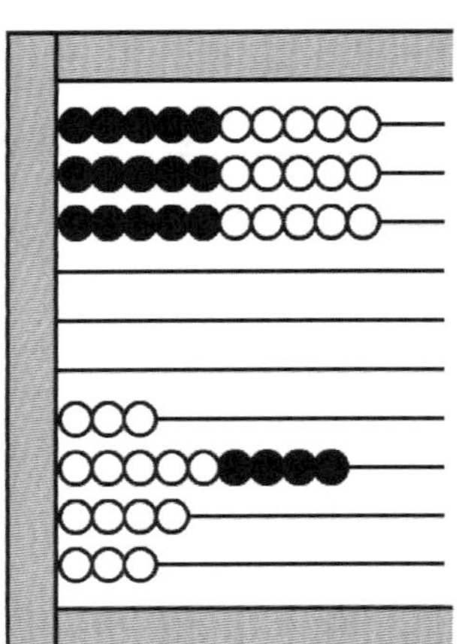

enter a ten on the second wire and enter a 1 on the fourth wire.

Start again with the 1 on the fourth wire; count the 4 on the fifth wire; and the 5 on the sixth wire. Since they equal 10, they are replaced with a ten.

Now start with the 3 on the seventh wire and continue counting on the eighth row as before. Clear and enter 1 ten and 2 ones.

This leaves 9 beads that can be cleared and entered on the fifth row. **So, what did all those numbers add up to be?** [49] Comment on the guesses.

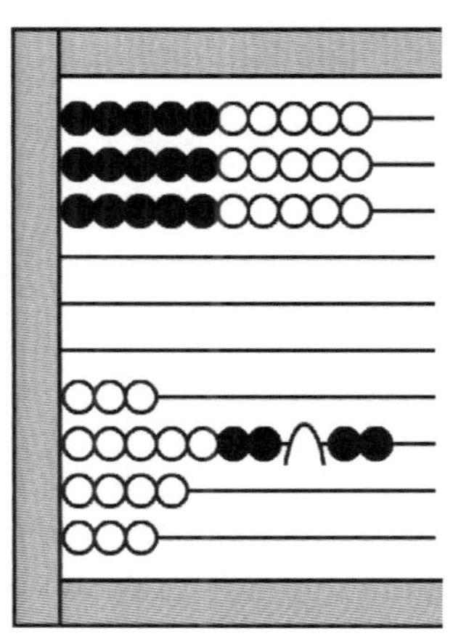 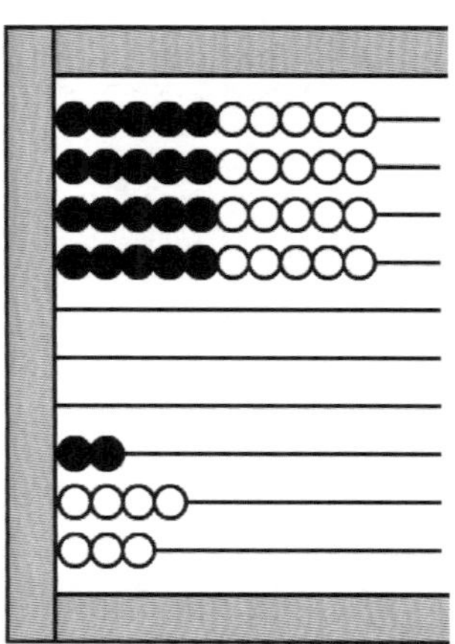 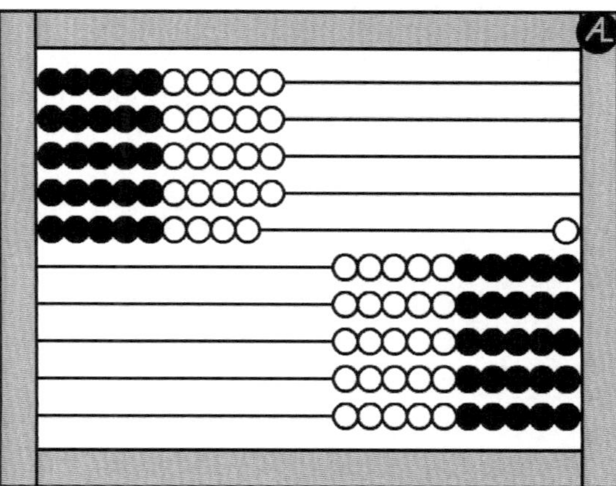

Guide the children through the process. Then give them another column of numbers to enter and add.

2. Alternately, provide them with ten or fewer cards, each having a number less than 10. When they obtain their first answer, ask them to write it down. Next they shuffle the ten cards and again add the numbers. Finally, they compare results.

Adding three numbers

Write on the board:

$$5 + 4 + 1 =$$
$$4 + 1 + 5 =$$
$$1 + 5 + 4 =$$

Show the children how to perform the additions by adding the first two numbers and then adding the third number. Choose three children to write down the sums.

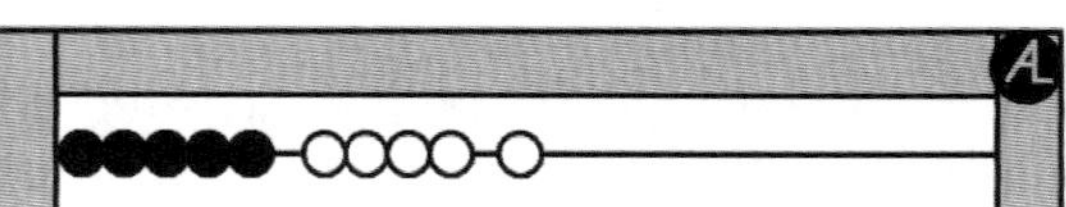

Repeat for the following set:

$$8 + 7 + 2 =$$
$$2 + 7 + 8 =$$
$$2 + 8 + 7 =$$

Have the children work in groups of three. Give each group three numbers with each member adding the numbers in a different order. Then they compare answers.

THOUGHT QUESTION: The answer is always the same no matter what order, but is one order easier than the others? [It is easier to form groups of 10.]

Give the children worksheets (3-34 and 3-35) similar to the one shown.

ORAL PROBLEMS. A. Jarvis collects baseball cards. He has 7 cards from the Yankees, 8 from the Cardinals, and 5 from the Twins. How many does he have from those three teams? [20 cards]

5	+	4	+	1	=	
6	+	1	+	5	=	
9	+	9	+	6	=	
3	+	8	+	4	=	
4	+	9	+	1	=	
8	+	1	+	2	=	
6	+	4	+	9	=	
10	+	5	+	3	=	
1	+	9	+	3	=	
6	+	0	+	6	=	
9	+	3	+	7	=	
4	+	6	+	2	=	

COUNTING TO 100

At this point, when the children understand the numbers to 100 and have memorized the customary names, it is time to work on counting to one hundred by ones, twos, fives, and tens.

Counting by ones

1. Since the children can name any quantity on the abacus, counting to one hundred is as simple as entering each bead, one at a time, row by row, and naming the quantity. The children can do this as a group, with partners, or individually.

To count beyond 100, use a second abacus. Be sure the children say "one hundred one," not "one hundred and one."

When they can do this easily with the abacus, challenge them to count without it.

2. WRITING THE HUNDRED CHART. Give the children paper (3-36) with ten rows of ten columns for writing to 100. As each bead is entered, the child forms it with the place-value cards and copies it on the paper. A calculator can be used either in conjunction with this work or to check it.

3. Provide the children with pennies. Show a PENNY and say, **This is a penny; it equals one CENT.** Let the children count the pennies; encourage them to arrange them in groups of tens. Each group of ten can be entered on the abacus.

When we count money and reach one hundred, it is called one DOLLAR. Show them a dollar bill. **One hundred cents equals what?** [one dollar]

4. Write

32 __ __ __

and ask, **What numbers come after 32?** [33 34 35] The children may need to use the abacuses. Repeat for other numbers: 23, 39, 84, and so on. This work can also be done on paper (3-37).

1									

32 __ __ __
92 __ __ __
70 __ __ __
26 __ __ __
58 __ __ __
47 __ __ __
69 __ __ __
11 __ __ __

Counting by 2s

1. Counting by 2s was introduced earlier. Now it can be extended to 100. Show the children how to move 2 beads at a time and name the quantity. If desired, a gap may be left between each group of two. Children may do this with partners and take turns with the rows.

To have the children write the results, give them a grid with ten rows of five rectangles (3-38). Also ask them to lightly color on the hundred chart the rectangles with even numbers.

2. Write

42 __ __ __

and ask, **What are the next even numbers?** [44 46 48] Repeat for 72, 38, and 40. Give similar written work (3-39). Allow use of the abacus, if needed.

3. Ask the children, **What do we call a number that is not even?** [odd] **What are the first five odd numbers?** [1, 3, 5, 7, 9] Ask them to refer to their colored hundred chart and read the odd numbers.

4. Say, **Since you know how to find the next even number, now could you find the next odd numbers?** Write

41 __ __ __

and ask, **What are the next odd numbers?** [43 45 47] Repeat for 13, 89, and 27. Give similar written work (3-40).

5. Show the children how to play an In-Order Memory. Use cards with numbers from 1 to 10. One player collects the evens in order while the other collects the odds in order.

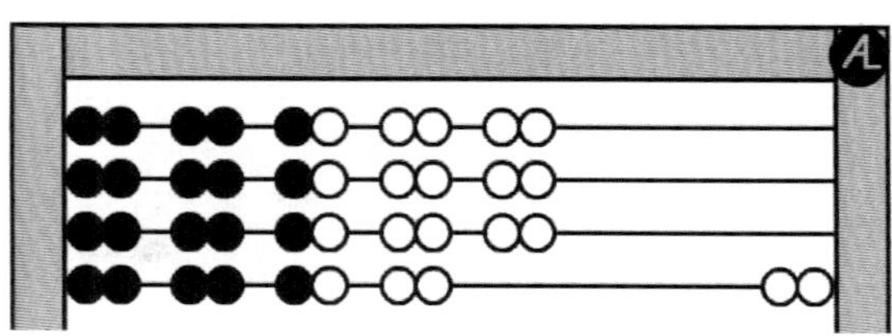

42	__ __ __
24	__ __ __
36	__ __ __
82	__ __ __
46	__ __ __
60	__ __ __
88	__ __ __
56	__ __ __

41	__ __ __
15	__ __ __
19	__ __ __
57	__ __ __
85	__ __ __
23	__ __ __
89	__ __ __
93	__ __ __

Counting by 10s

Counting by tens was introduced earlier. Review or introduce a DIME. Explain that a dime equals 10 cents, so they are counted by 10s; for example, 10 cents, 20 cents, and so forth. Give the children several dimes to count. Remind them that 100 cents equals one dollar.

Give them various combinations of dimes and pennies and ask them to determine the value.

Counting by 5s

1. Memorizing the 5s is necessary for learning money and telling time. To count by 5s, move over 5 beads at a time while the children recite the quantities. Have the children count by 5s with a partner. To write the 5s on their own, the children need a grid (3-38) as shown.

2. After the 5s are mastered, show the children a nickel and tell them it is worth 5 cents and ask, **How do we count nickels?** [by 5s] Show them several nickels and ask, **How much are these nickels worth?** Initially, the children can enter on the abacus 1 five for each nickel. Later, challenge them to touch each nickel in turn while reciting the 5s.

As review ask, **How many pennies are needed to make a dollar?** [100]

THOUGHT QUESTION: How many dimes are needed to make a dollar? [10]

THOUGHT QUESTION: How many nickels are needed to make a dollar? Give the children time to figure out the answer. [20]

3. Give them several nickels and pennies and ask them count by 5s and 1s to determine how much it is. Also include dimes for them to count.

4. ORAL PROBLEMS. A. Andy has 60 cents and earns a nickel. How much money does he have now? [65 cents]

B. Robyn brought 4 trucks into the tent; Chris brought 7 cars; and Billie brought 6 balloons. How many toys were in the tent? [17 toys]

C. Jessica has 6 outfits for her doll. A friend gave her 2 more. How many does she have now? [8 outfits]

D. Kevin has 6 nickels. How much money is this? [30 cents]

E. Joni has 4 dimes and 1 penny. How much money is this? [41 cents]

F. Grandma Anderson has 9 granddaughters and 6 grandsons. How many grandchildren does she have? [15 grandchildren]

G. Last August Cheri saw 4 shooting stars on Monday, 6 on Tuesday, and 5 on Wednesday. How many did she see those three days? [15 shooting stars]

H. Dawn rode 32 miles to visit her grandfather and 32 miles back home. How many miles did she ride both ways? [64 miles]

I. Carla is counting the minutes on a clock. Each large number means 5 minutes. How many minutes are there between the 4 and the 8? [20 minutes]

J. Miles walked up three flights of stairs. The first flight had 19 steps; the second and third flights had 20 steps each. How many steps did he walk up all together? [59 steps]

Unit 4
Mastering the addition facts

This unit teaches strategies for mastering the addition facts. A strategy allows the children to mentally construct the facts. Children gain in self-esteem and independence because they feel competent at reconstructing a forgotten fact. Strategies, which are first concretely taught on the abacus, involve thinking skills that have other applications. Using flash card drill to teach the facts without providing strategies requires only rote memory, the lowest form of thinking.

The strategies for the addition facts form four clusters: adding 1 or 2; doubles and variations; facts that equal 10, 11 or 9; and adding 9 or 8. (Incidentally, it is not true that children learn the doubles easier than other facts. Children know them better because teachers and parents drill them more.) Every addition fact is included at least once in these clusters. The accompanying addition tables show the facts included in each cluster. It is helpful for the children to color on an addition table (see the appendix) the facts they have learned. To allow more practice in using the strategy, the suggested practice often includes work in the higher decades.

Although important, strategies are insufficient to master the desired quick recall. The learner must also practice through interesting repetition. For games stressing these clusters of strategies and other practice, see *Math Card Games* by the same author.

+	1	2	3	4	5	6	7	8	9
1	2	3	4	5	6	7	8	9	10
2	3	4	5	6	7	8	9	10	11
3	4	5	6	7	8	9	10	11	12
4	5	6	7	8	9	10	11	12	13
5	6	7	8	9	10	11	12	13	14
6	7	8	9	10	11	12	13	14	15
7	8	9	10	11	12	13	14	15	16
8	9	10	11	12	13	14	15	16	17
9	10	11	12	13	14	15	16	17	18

Facts +1 and +2.

+	1	2	3	4	5	6	7	8	9
1	2	3	4	5	6	7	8	9	10
2	3	4	5	6	7	8	9	10	11
3	4	5	6	7	8	9	10	11	12
4	5	6	7	8	9	10	11	12	13
5	6	7	8	9	10	11	12	13	14
6	7	8	9	10	11	12	13	14	15
7	8	9	10	11	12	13	14	15	16
8	9	10	11	12	13	14	15	16	17
9	10	11	12	13	14	15	16	17	18

Facts =10, =11, and =9.

+	1	2	3	4	5	6	7	8	9
1	2	3	4	5	6	7	8	9	10
2	3	4	5	6	7	8	9	10	11
3	4	5	6	7	8	9	10	11	12
4	5	6	7	8	9	10	11	12	13
5	6	7	8	9	10	11	12	13	14
6	7	8	9	10	11	12	13	14	15
7	8	9	10	11	12	13	14	15	16
8	9	10	11	12	13	14	15	16	17
9	10	11	12	13	14	15	16	17	18

Facts with doubles and variations.

+	1	2	3	4	5	6	7	8	9
1	2	3	4	5	6	7	8	9	10
2	3	4	5	6	7	8	9	10	11
3	4	5	6	7	8	9	10	11	12
4	5	6	7	8	9	10	11	12	13
5	6	7	8	9	10	11	12	13	14
6	7	8	9	10	11	12	13	14	15
7	8	9	10	11	12	13	14	15	16
8	9	10	11	12	13	14	15	16	17
9	10	11	12	13	14	15	16	17	18

Facts +9 and +8.

+	1	2	3	4	5	6	7	8	9
1	2	3	4	5	6	7	8	9	10
2	3	4	5	6	7	8	9	10	11
3	4	5	6	7	8	9	10	11	12
4	5	6	7	8	9	10	11	12	13
5	6	7	8	9	10	11	12	13	14
6	7	8	9	10	11	12	13	14	15
7	8	9	10	11	12	13	14	15	16
8	9	10	11	12	13	14	15	16	17
9	10	11	12	13	14	15	16	17	18

Facts using the 2-fives strategy.

EVEN + EVEN...

A study showed that a majority of arithmetic errors in addition and subtraction were only one number off. But a person aware of the following observations cannot be one number off:

> Even + Even = Even
> Odd + Odd = Even
> Even + Odd = Odd

Therefore, the object of this lesson is to help the children discover and apply the rules governing the sum of two evens, two odds, or one of each.

Even + even

1. Review the even numbers by first asking the children to say the even numbers up to 20. **When we say the even numbers, we are also doing something else. What is it?** [counting by 2s] Then ask, **Is 7 an even number?** [no] **Is 2 an even number?** [yes] **Is 12 an even number?** [yes]

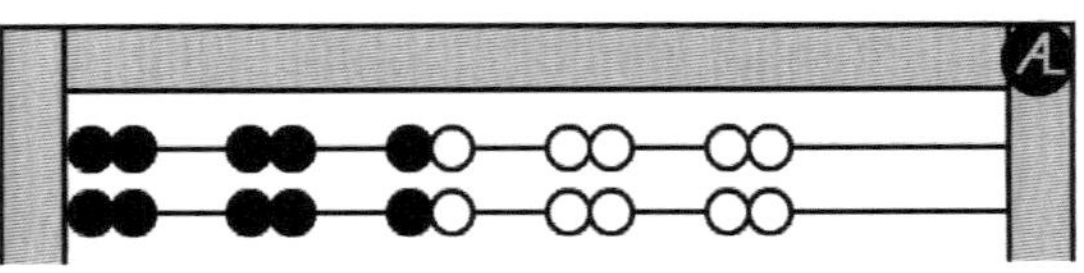

Now let's see what happens when we add two even numbers. Use the vertical format on the abacus. Write

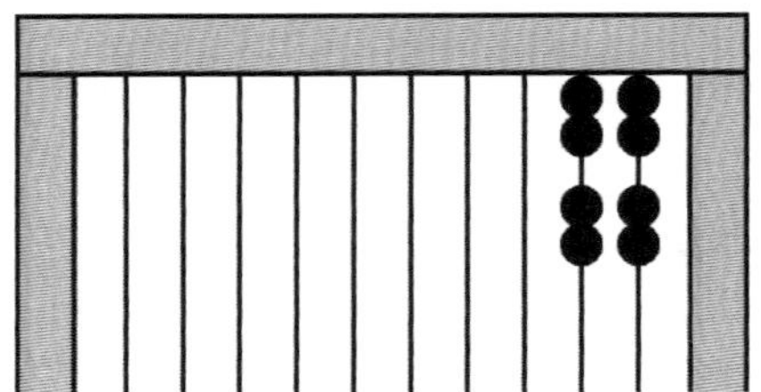

$$\begin{array}{cccc} 4 & 6 & 8 & 4 \\ +4 & +2 & +6 & +8 \end{array}$$

and say, **Add 4 plus 4; is the sum an even number?** [yes] **What about 8 + 2?** [even] Repeat for 6 + 6 and 4 + 8. **Can you think of two even numbers we can try?** Write the responses on the board and ask the children to add them.

2. Give the children a blank paper (3-33) for vertical addition. Ask them to think of even numbers no higher than 10, to write them, and then to add them. Also tell them to mark the even sums in some way, such as circling or coloring when the page is complete. Advanced children may try numbers greater than 10, using the abacus in the horizontal format.

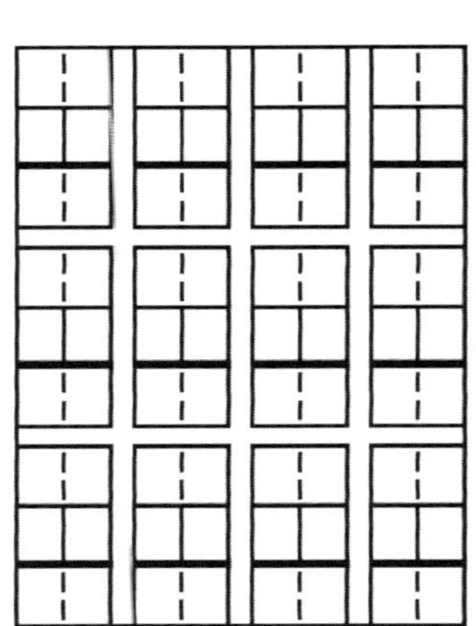

The children can also work with partners, each thinking of an even number. They can take turns adding them and writing the sum.

Odd + odd

1. Start by saying, **What happened when we added two even numbers?** [The sum is even.] **Now let's see what happens when we add two odd numbers?** Ask several children what their guesses are. **Now let's find out.**

Review by asking the children to say the odd numbers up to 20. Use the vertical format on the abacus and the board. Write,

$$\begin{array}{cccc} 5 & 7 & 5 & 7 \\ +\,3 & +\,1 & +\,5 & +\,9 \end{array}$$

Ask the children to add 5 plus 3; is the sum an odd number? [no] **What about 7 + 1?** [even] Repeat for 5 + 5 and 7 + 9. **Can you think of two odd numbers we can add?**

2. Give the children a blank vertical addition sheet to write their own odd numbers and the sums. Even answers should be marked in some fashion, by circling or coloring. (Obviously, there will not be any.)

They could work with partners, as before.

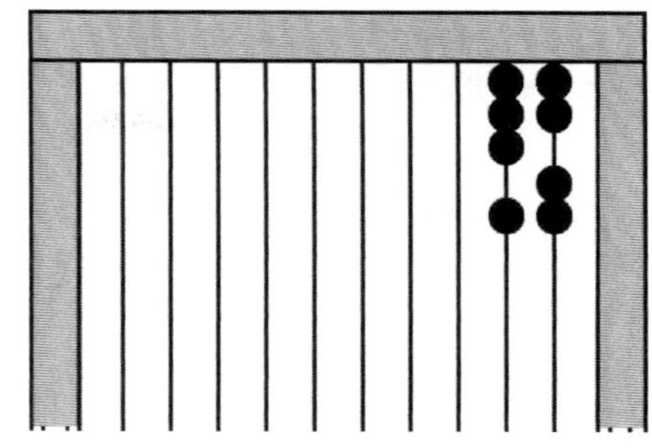

Even + odd

Review by asking, **What do you get when you add two even numbers?** [even number] **What do you get when you add two odd numbers?** [even number] **What do you think you would get if one number is odd and one is even?**

Write and say,

$$\begin{array}{c} 4 \\ +\,5 \end{array}$$

Add 4 plus 5; what is the sum, even or odd? [odd] **What about 3 + 2?** [odd] Write on the board

$$\begin{array}{c} 9 \\ +\, \end{array}$$

Let the children think of an appropriate number to add. Repeat for

$$\begin{array}{c} 6 \\ +\, \end{array}$$

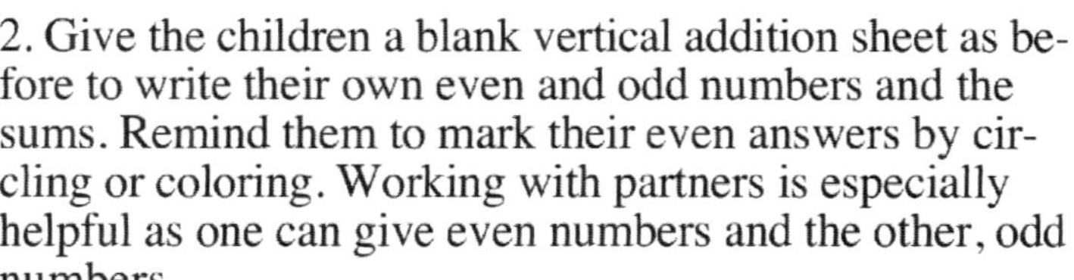

2. Give the children a blank vertical addition sheet as before to write their own even and odd numbers and the sums. Remind them to mark their even answers by circling or coloring. Working with partners is especially helpful as one can give even numbers and the other, odd numbers.

3. Give a worksheet in the vertical format and ask the children to mark all the sums that will be even. You might then want them to check their work by actually performing the addition. Previous worksheets in either format could also be used.

4. After the children have completed the worksheets (4-1), ask them if they could name a simple way to remember whether the sum will be even or odd. [One possible statement: the sum is even only if both numbers are even or both numbers are odd.]

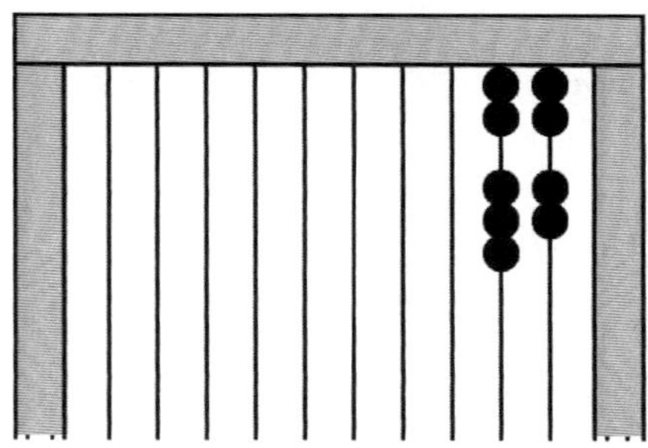

ADDING 1 OR 2

Earlier, the children were guided in discovering that adding 1 to a number resulted in the next higher number. This concept will be reviewed and expanded to higher decades.

Adding 2 to a number will be divided into two strategies. Adding 2 to an even number results in the next higher even number; whereas, adding 2 to an odd number gives the next higher odd number.

Adding 1 to a number

Ask the children to build the stairs. Then say, **Touch the wire with 4; what number comes after 4?** [5] **Touch the wire that comes after it.** Repeat for wires with 6 and 2.

Now find the wire with 4 again and add 1 to it. What does it equal? [5] **Find the wire with 7 and add 1 to it; what is the sum?** [8] **So, what happens when we add 1 to a number?** [The sum is the next higher number.]

Expand by asking what number comes after 25, 38, and so forth. Then give the children various problems, both written on the board and orally, involving adding 1; for example, 9 + 1, 13 + 1, 22 + 1, 67 + 1, 99 + 1?

Next ask, **What is 1 + 23?** [24] **How could you figure it out?** [It is the same as 23 + 1.] Practice with 1 + 8, 1 + 14, 1 + 87 before giving them worksheets (4-2) for written work.

THOUGHT QUESTION: If you add 1 to an even number, will the sum be even or odd? [odd]

THOUGHT QUESTION: If you add 1 to an odd number, will the sum be even or odd? [even]

Adding 2 to a number

This activity demonstrates adding 2 to a number by the process of counting up 2. Another method, somewhat more sophisticated and faster, uses the concept of even and odd and is discussed next. Ask the children to build the stairs. Tell them to add 2 to the 3 and ask, **Which number does it look like now?** [5] **How far away is it?** [Two below]

Repeat with adding 2 to the 6 and adding 2 to 7. Emphasize that in each case the sum is the same as counting 2 more, or the same as adding 1 two times.

Adding 2 to an even number

1. Review the even numbers, 2, 4, 6, 8, 10. Show the children how to make the even stairs with 2 on the first wire, 4 on the second wire, and so forth. Ask, **What is the next even number after 8?** [10] **What is the even number following 4?** [6]

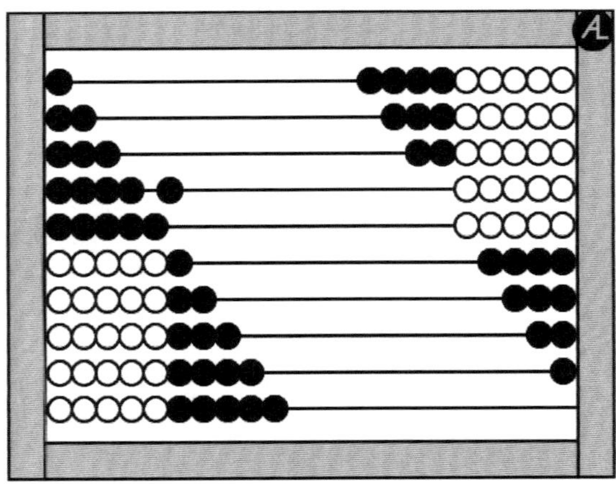

47	+	1	=	
85	+	1	=	
54	+	1	=	
79	+	1	=	
62	+	1	=	
1	+	98	=	
1	+	26	=	
1	+	13	=	
1	+	31	=	

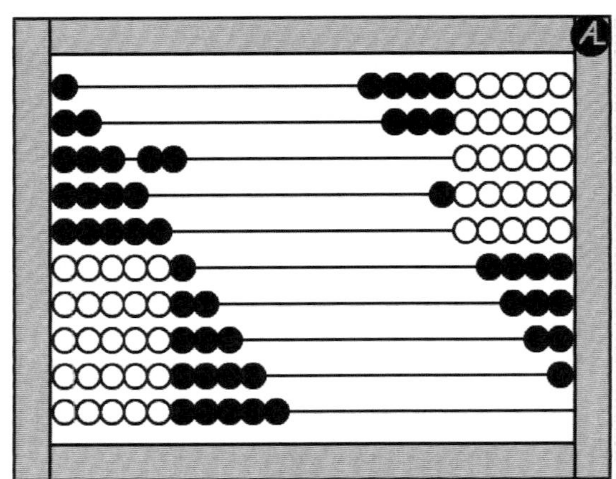

Add 2 to the wire with 8; what does it equal?
[10] **Add 2 to the 4; what does it equal?** [6] **So,
to add 2 to an number, the rule is...** [The sum is
the next higher even number.]

2. With the wires vertical, ask the children to enter 2 and
to add 2 and recite the fact. [2 + 2 = 4] Ask them to add
another 2 and recite that fact. [4 + 2 = 6] Continue to 10.

This is a good time to review the commutative law, but
now with vertical wires. For comparison purposes, the
sixth and seventh wires will be used as a set. In the future, those two wires will be used to show the tens place.

Show the children how to enter 8 + 2 and 2 + 8 as
shown. Ask a child if they look the same. [yes] They
might notice that one is upside down from the other. To
see this more clearly one child lays her abacus upside
down next to another right side up.

Give them a worksheet (4-3) similar to the one shown.

ORAL PROBLEMS. A. Krystal's family refrigerator had
6 drawings on it. She brought home 2 more and taped
them up also. How many drawings are on the refrigerator
now? [8 drawings]

B. Mike had 6 balls and bought 2 more. Brian had 2 balls
and bought 6 more. Who now had more balls? [They had
the same.]

3. As review, ask the children to recite the even numbers
to 100. Then ask, **What even number comes after
24?** [26] Write and say 24 + 2 = . Encourage the children
to guess and then check their answers on the abacus. Repeat with other numbers, such as 36 + 2, 42 + 2, and 58
+ 2, but first ask them to name the next higher even number. Then ask, **What is the rule for adding 2 to an
even number.** [The next even number]

Challenge the children with the problem 2 + 26. Let them
find the solution on the abacus any way they can. If necessary say, **Does it matter which number we start
with?** [no] and suggest that they start with the 26 and
then add 2. Give them other problems, such as 2 + 82, 2
+ 46, and 2 + 88.

When the children are confident with the rule, give them
worksheets (4-4) to be done without the abacus.

Adding 2 to an odd number

1. Be sure the children our very secure with adding 2 to
an even number before starting this activity. Begin by reviewing the odd numbers to 99. Make a game of saying
or writing an odd number and the children naming the
next higher odd number. Further involve the children by
asking one child to name the first number and another
child to name the next odd number.

Ask the children to build the odd stairs. Say, **What happens if I add 2 to the 3; what row does it equal?**

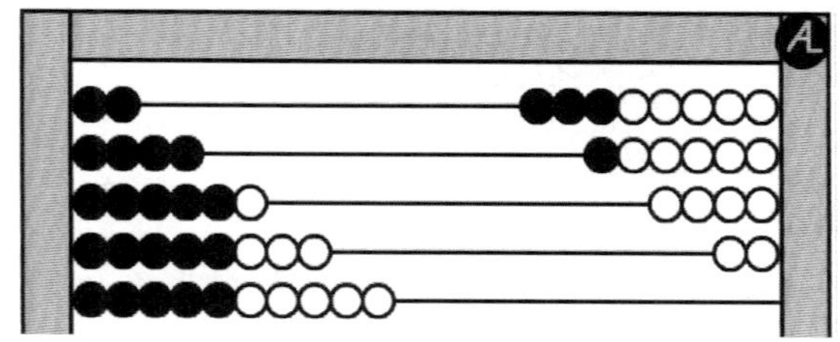

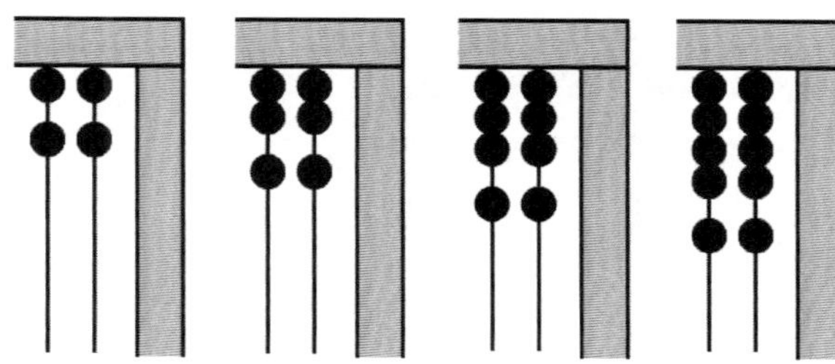

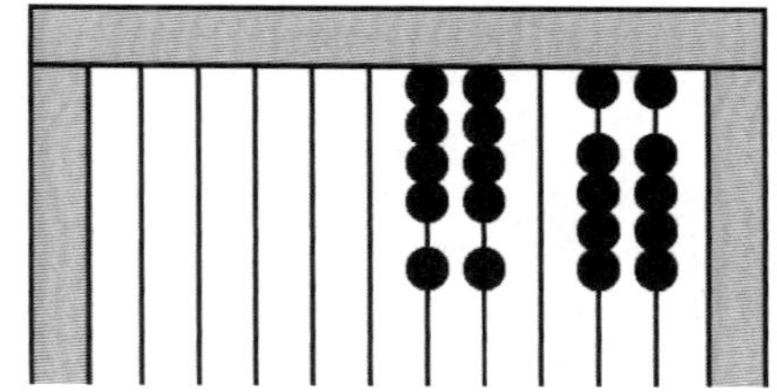

4	2	0	2		24	+	2	=	
+2	+2	+0	+8		30	+	2	=	
					76	+	2	=	
6	0	2	2		94	+	2	=	
+2	+2	+2	+6		82	+	2	=	
					2	+	64	=	
2	2	2	8		2	+	34	=	
+0	+6	+4	+2		2	+	58	=	
					2	+	16	=	

Add 2 to the 3. [It is equal to 5, the same as the row below.] Repeat with 5 + 2 and 7 + 2. **What happens when we add 2 to an odd number?** [The next odd number]

2. With the wires vertical, ask the children to enter 1, then to add 2 and recite the fact. [1 + 2 = 3] Continue to 9 + 2 as shown below. Also provide them with a worksheet (4-5) as shown.

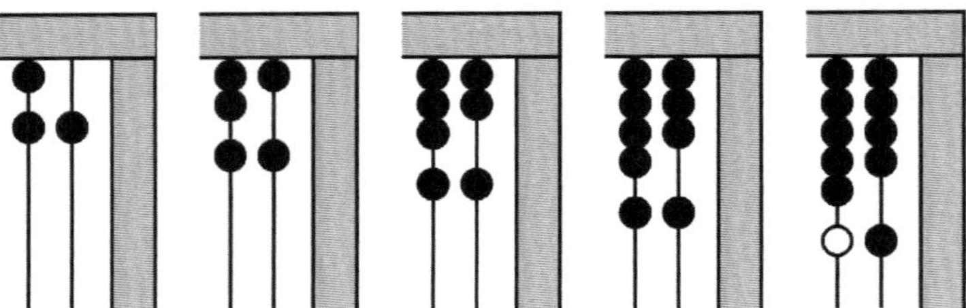

Continue the lesson with two digit numbers, such 31 + 2, 73 + 2, 29 + 2 and 2 + 12, 2 + 35, 2 + 81. Provide worksheets (4-6) as shown.

3. Finally, write on the board a mixture of even and odd numbers and ask the children to add 2. Use, for example, 80 + 2, 57 + 2, 35 + 2, and 28 + 2. The abacus should not be needed at this point. Give them worksheets (4-7).

Give the children a mixture (4-8) of two-digit numbers + 1 and + 2 or a page (4-9) of the basic facts involving + 1 or + 2.

SUMS EQUAL TO 11 OR 9

The objective is that the children will correlate the facts totaling 11 or 9 with those totaling 10. If either of the two addends totaling 10 is increased by 1, the sum will be 11; if decreased by 1, the sum will be 9.

Facts equal to 11

Before approaching the facts equaling 11, explore the idea of what happens when we add 1 to one addend.

1. Write for the children to see

$$6 + 3 =$$
$$6 + 4 =$$

Ask them to find partners. One enters the first sum on the top wire and the other enters the second sum on the second wire. Call on various children to write the sums. Repeat for 2 + 7 and 2 + 8 and for 7 + 5 and 7 + 6. This last case requires two wires for each sum. Then ask, **What happens to the sum if one of the ADDENDS is one greater?** [The sum is one greater.]

If desired, give the children a worksheet (4-10) similar to that shown.

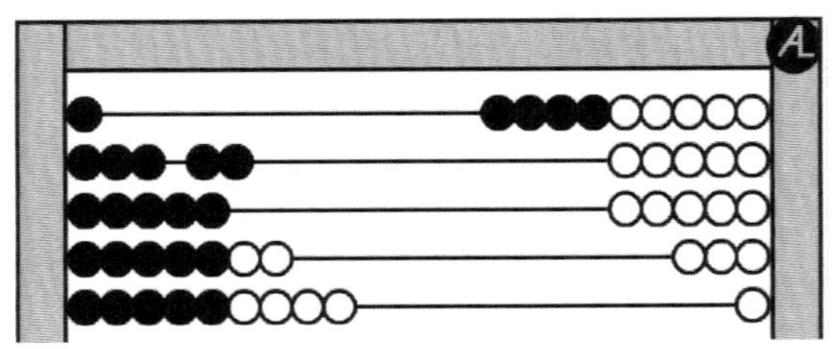

3	2	5	1
+2	+1	+2	+2

7	2	9	2
+2	+3	+2	+7

2	5	2	2
+9	+2	+7	+5

31	+	2	=	
37	+	2	=	
65	+	2	=	
79	+	2	=	
83	+	2	=	
2	+	57	=	
2	+	43	=	
2	+	69	=	
2	+	81	=	

80	+	2	=	
77	+	2	=	
59	+	2	=	
18	+	2	=	
26	+	2	=	
81	+	2	=	
34	+	2	=	
63	+	2	=	
45	+	2	=	

8	+	1	=	
5	+	2	=	
9	+	1	=	
4	+	2	=	
6	+	1	=	
5	+	1	=	
1	+	2	=	
8	+	2	=	
2	+	1	=	

6	+	3	=	
6	+	4	=	
2	+	6	=	
2	+	7	=	
8	+	5	=	
8	+	6	=	
7	+	2	=	
7	+	3	=	

2. Enter 10 on the top wire; skip a wire and enter 11 as shown. Separate the last bead of the 10 and ask a child to state the fact and to write it for all to see. [9 + 1 = 10] Likewise, separate a bead from the 11 and ask, **N o w what fact do we have?** [9 + 2 = 11] Call on someone to write it below the first fact.

Continue with the other pairs of facts down to 2 + 8 and 2 + 9. You might ask the children to write all the facts equaling 11 on blank squared paper. Give them worksheets (4-11and 4-12).

ORAL PROBLEMS. A. Mitchell and his friend Tyler had 10 transformers. How many different ways could they be split between the boys? [11: 0 and 10, 1 and 9, and so forth] Have the children make tables showing the possibilities:

Mitchell	Tyler
0	10
1	9
2	8

B. Bethany and Laura had 9 dinosaurs. How many different ways could they be split between them? [10]

Facts equal to 9

The strategy for facts equaling 9 is very similar to that of facts equaling 11. For that reason, be sure the children are very secure with the 11s before proceeding to the 9s.

Ask the children to enter 10 on the top wire and 9 on the second wire. Separate 2 beads on the top wire and 1 bead on the second wire as shown. Call on various children to state the two facts. [8 + 2 = 10 and 8 + 1 = 9] The pair of facts may be recorded on the chalkboard.

Move over another bead on each wires and ask for the facts. [7 + 3 = 10 and 7 + 2 = 9] Continue until only 1 bead remains.

ORAL PRACTICE. Ask the children to recite all the ways to make 11. [9 + 2, 8 + 3, ... 2 + 9]

For written practice, the children could write the facts equaling 9, starting with 8 + 1. Then give them worksheets (4-13 and 4-14).

Two-fives

There is a simple strategy for adding two numbers when both numbers are between 5 and 10. Enter 7 and 8 on the wires of the abacus as shown. **Can you see the 10?** [the 2 groups of 5 beads, the dark-colored beads] **H o w much is left over?** [the 3 and 2, or 5] **How much is 7 + 8?** [15] Ask them to see it in their minds.

Continue with other facts such as 5 + 7, 6 + 7, 8 + 5, 7 + 9, 5 + 9, and 5 + 6.

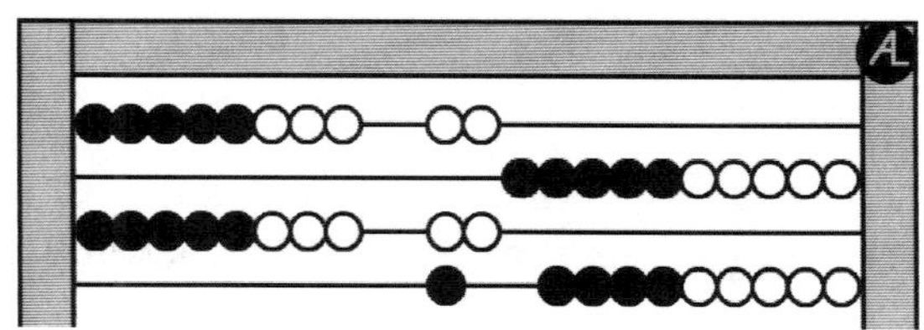

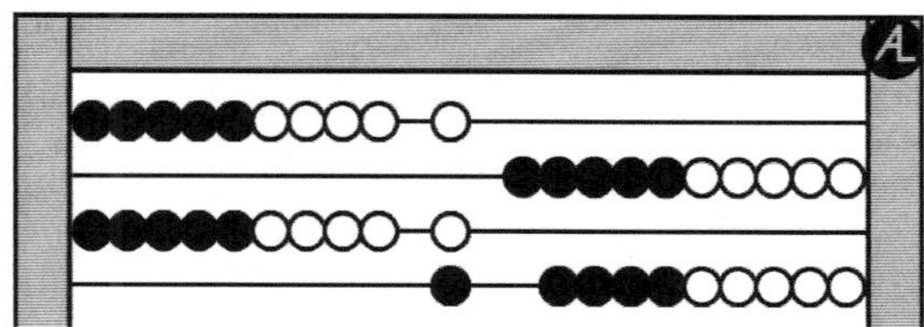

8	+		=	10
8	+		=	11
3	+		=	10
3	+		=	11
6	+		=	10
6	+		=	11
5	+		=	10
5	+		=	11

3	+	7	=	
8	+	3	=	
7	+	4	=	
9	+	2	=	
4	+	6	=	
5	+	6	=	
4	+	7	=	
8	+	2	=	
6	+	5	=	

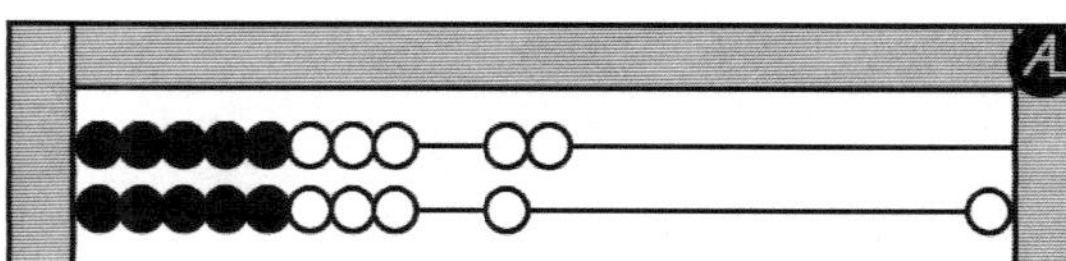

5	+		=	10
5	+		=	9
2	+		=	10
2	+		=	9
8	+		=	10
8	+		=	9
3	+		=	10
3	+		=	9

7	+	3	=	
4	+	5	=	
1	+	8	=	
2	+	8	=	
4	+	7	=	
5	+	4	=	
6	+	3	=	
3	+	8	=	
7	+	4	=	

THE DOUBLES GROUPS

Besides the doubles (3 + 3), this group contains the near doubles (3 + 2) and the middle doubles (3 + 5). The doubles groups includes 48% of the addition facts.

Doubles

A simple way of visualizing the doubles on the abacus is to enter the quantities with the wires horizontal and read the sum with the wires vertical.

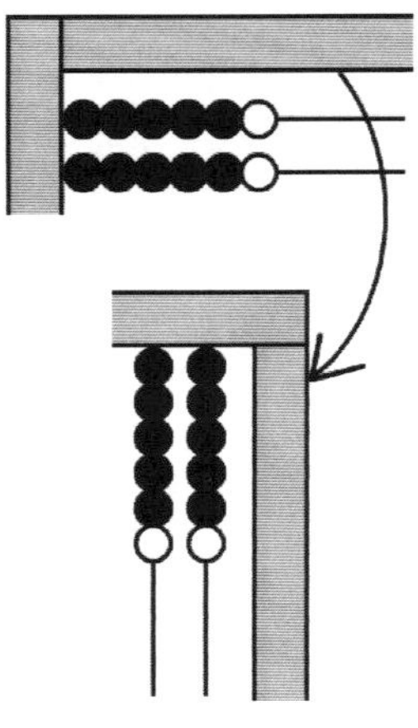

Write on the board the doubles:

$$1 \quad 2 \quad 3 \quad 4 \quad 5 \quad 6 \quad 7 \quad 8 \quad 9$$
$$\underline{1} \quad \underline{2} \quad \underline{3} \quad \underline{4} \quad \underline{5} \quad \underline{6} \quad \underline{7} \quad \underline{8} \quad \underline{9}$$

Tell the children that these facts are called the doubles and that today they will learn a way to remember them. Ask the children to enter 1 on the top two wires and then to turn it and read the sum. Call upon a child to recite the entire fact [One plus one is two.] and write the sum.

Continue with the remaining doubles. Note that the units repeat starting with 6.

$$1 \quad 2 \quad 3 \quad 4 \quad 5 \quad 6 \quad 7 \quad 8 \quad 9$$
$$\underline{1} \quad \underline{2} \quad \underline{3} \quad \underline{4} \quad \underline{5} \quad \underline{6} \quad \underline{7} \quad \underline{8} \quad \underline{9}$$
$$2 \quad 4 \quad 6 \quad 8 \quad 10 \quad 12 \quad 14 \quad 16 \quad 18$$

Ask the children if they notice anything special about the sums. [They are the even numbers.]

ORAL PRACTICE. Have the children practice with partners reciting the doubles. With the wires vertical, the children enter 2 in the even-odd format and state $1 + 1 = 2$. Next they enter 1 more on each wire and recite $2 + 2 = 4$. Continue to $9 + 9$.

If desired, give the children blank pages for recording the doubles on their own. Also give them worksheets (4-15 and 4-16) as shown.

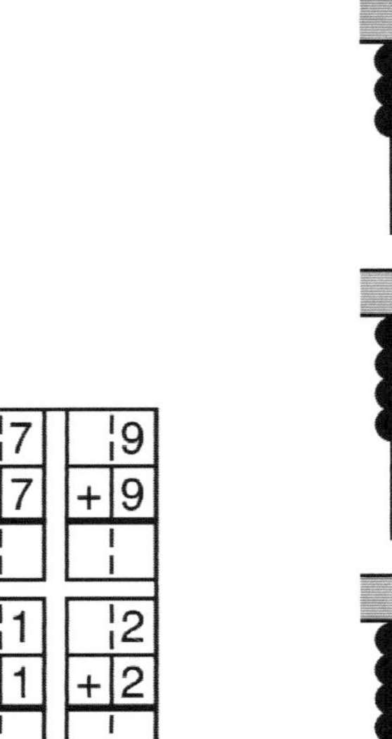

6		3		7		9	
+6		+3		+7		+9	
5		8		1		2	
+5		+8		+1		+2	
8		4		9		8	
+8		+4		+9		+8	

Near doubles

Be certain the doubles are thoroughly mastered before working on the near doubles because the strategy for learning the near doubles is the double plus 1.

Write on the board the near doubles:

$$1 \quad 2 \quad 3 \quad 4 \quad 5 \quad 6 \quad 7 \quad 8$$
$$\underline{2} \quad \underline{3} \quad \underline{4} \quad \underline{5} \quad \underline{6} \quad \underline{7} \quad \underline{8} \quad \underline{9}$$

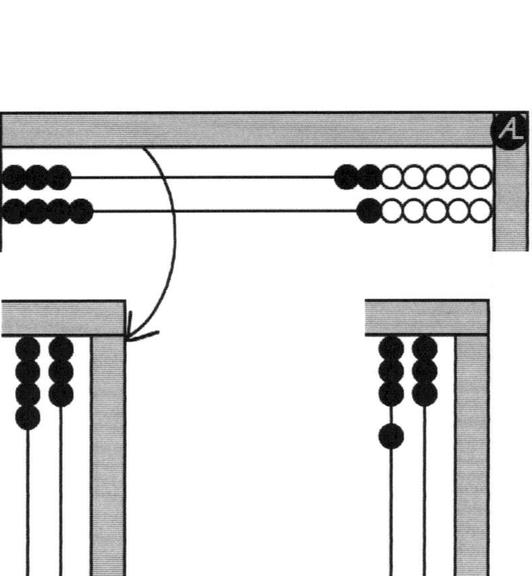

Tell the children that now they are going to work on the near doubles. On your abacus enter 3 on the top wire and 4 on the next wire. Use your thumb to keep the beads from falling and turn the abacus sideways. Slide the odd bead down an short distance and ask, **3 + 4 looks like what double.** [3 + 3] **S o, 3 + 4 equals what?** [7]

Repeat for 8 + 7. Help the children formulate the rule: to add near doubles, think of the double and add 1.

ORAL PRACTICE. With the wires vertical, ask the children to build the stairs up to 9. Then ask them to touch the first two wires and recite the fact. [1 + 2 = 3] Next they move their fingers over a wire and recite that fact. [2 + 3 = 5] Continue to 8 + 9.

Give the children practice sheets (4-17 and 4-18).

ORAL PROBLEMS. A. Stephie and Nicole are sisters. Each has 9 tee shirts. How many do they have together? [18 shirts]

B. Kyle can play 8 pieces and Peter can play 9 different pieces. How many different pieces could they play in a concert? [17 pieces]

C. Jo has 2 letters in her name. Ronny has 5 letters in his name. Melissa has 7 letters in her name. How many letters do they have altogether? [14 letters]

D. Dan worked 5 hours building a toy boat. His dad helped him and worked 2 hours. How many hours did it take to build the boat? [7 hours]

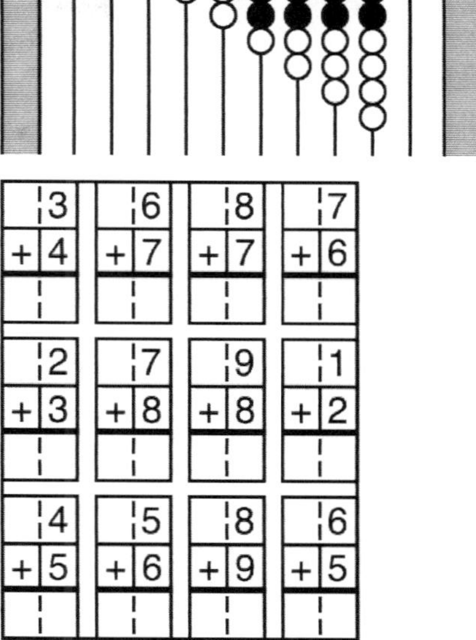

Middle doubles

The strategy for finding the sum when one number is 2 more than the other number is to find the middle and then think of its double.

Write

$$\begin{array}{ccc} 5 & 5 & 6 \\ +7 & +3 & +8 \end{array}$$

and say, **Now I will show you how to remember these kinds of facts. Enter 5 and 7 on the right two wires. To make them look like a double, I will TAKE off an extra bead and GIVE it to the other wire.** In one smooth motion, remove the last bead from the wire with 7 and add a bead to the wire with 5.

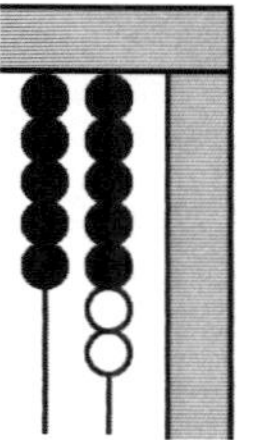

Do we still have the same amount? [yes] **What is the middle of 5 and 7.** [6] **What is the double of 6?** [12] **So how much is 5 + 7?** [12]

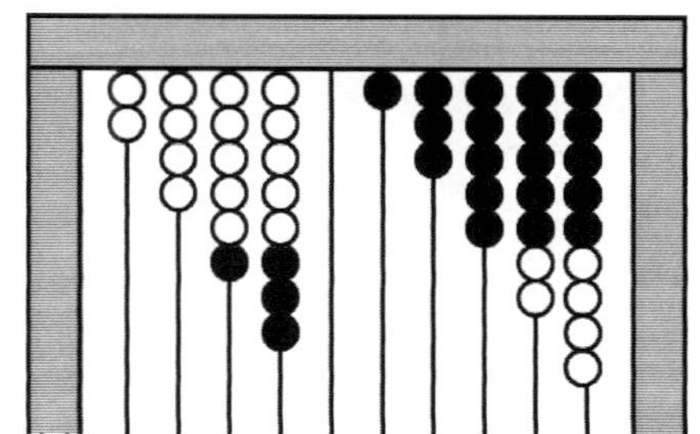

Repeat for 5 + 3 and 8 + 6. Help the children formulate the rule that to add a number to a number that is 2 higher, find the middle and think of its double.

ORAL PRACTICE. Ask the children to build the even stairs, skip a wire, and build the odd stairs as shown. Then ask them to touch the left two wires and say the fact. [1 + 3 = 4] Next they move to the next two wires and say that fact. [3 + 5 = 8] When the odds are completed, continue with the evens.

FUTURE REVIEW. Later, give them the facts at random and tell them to only look at their abacuses to find the answers.

Give them worksheets (4-19 and 4-20) for practice. If any children are having problems, suggest they write the middle numbers outside the boxes.

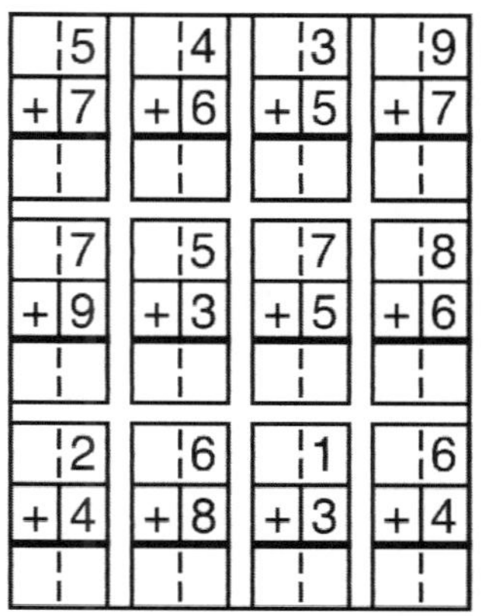

Also give them worksheets (4-21 and 4-22) in both formats with examples of all three types of doubles problems. At this point, the abacus should not be needed, but some children may be helped by the suggestion they first find and write the doubles sums, next they find all the near doubles, and finally do the middle doubles.

6	+	5	=	
8	+	9	=	
7	+	9	=	
3	+	3	=	
1	+	3	=	
8	+	8	=	
9	+	7	=	
4	+	6	=	
5	+	4	=	

ADDING NINE OR EIGHT

The strategy for adding 9 or 8 to a number is based on adding 10 to a number, which needs to be thoroughly understood first.

Adding 10 to a number

Write and ask the children to add 25 + 10; ask one of them to record the sum. [35]

Now I will show you a simpler way. We will enter the tens at the top of the abacus and the ones near the bottom. Show them how to enter the 25: the 2 tens at the top and the 5 ones on the second wire from the bottom. Then ask, **Where will we enter the 10 to be added?** [Next to the other tens.] **What is the sum?** [35] **Is the answer the same.** [yes]

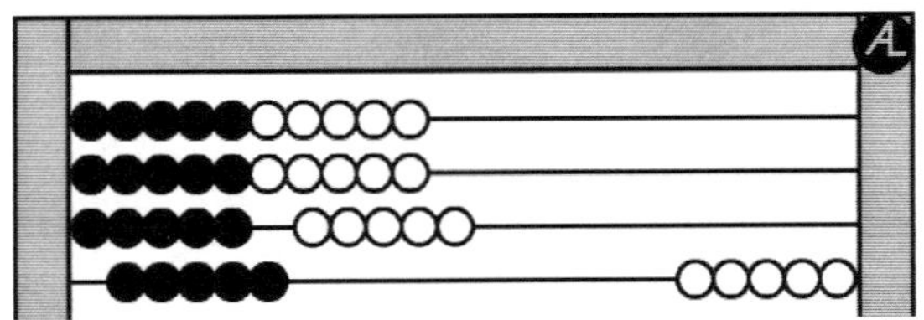

ORAL PRACTICE. Give the children numbers such as 43 + 10, 27 + 10, 38 + 10, and 19 + 10. Avoid numbers higher than the 60s. Ask, **What is the rule to add 10 to a number?** [The tens increase by 1 while the units remain the same.]

Write on the board the following:

39	53	42	14	21	35
+ 10	+ 10	+ 10	+ 10	+ 10	+ 10

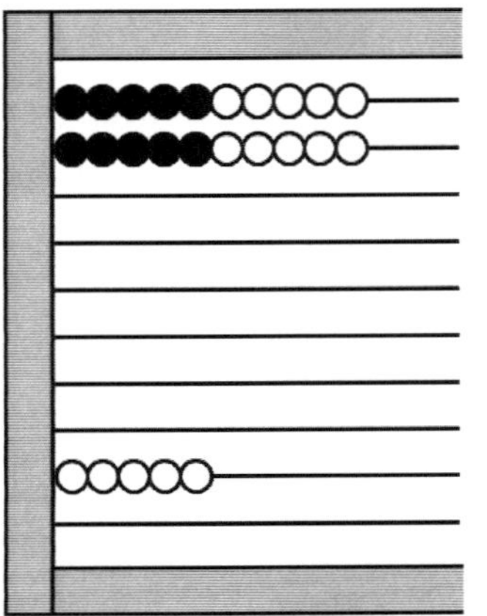

and tell the children that even though the problems are written in columns, they can be done the same way. Call upon various children to write the answers below the problems.

Provide worksheets (4-23 and 4-24) as shown. Some children may want abacuses to check their work.

Adding 9 to a number

Once adding 10 to a number is mastered, adding 9 can be taught. Adding 9 is the same as going back 1 (subtracting 1) and then adding 10.

As a quick review, ask the children to say the number before 8, [7] before 3, [2] before 6 [5].

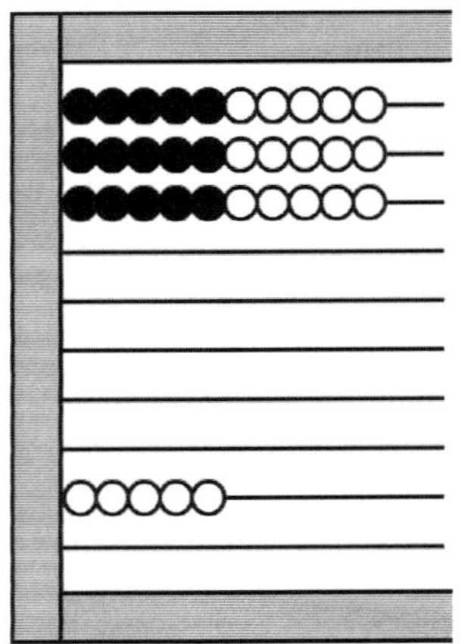

25	+	10	=	
49	+	10	=	
17	+	10	=	
56	+	10	=	
48	+	10	=	
33	+	10	=	
59	+	10	=	
22	+	10	=	
46	+	10	=	

Write on the board

3	9	7	9	4
+ 9	+ 5	+ 9	+ 6	+ 9

Tell them that today they will learn an easy way to remember the facts with 9. Ask them to enter 3 on the top wire and 9 on the next wire. Next show them how to Take 1 from the 3 and Give it the 9, making it 2 and 10. Then ask, **Now what is the answer?** [12]

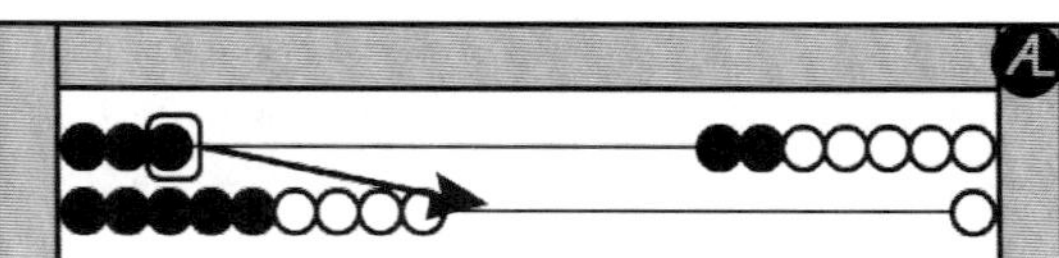

Repeat for the other facts. When the children understand the process of removing 1 from the other number, give them more oral examples and a worksheet (4-25).

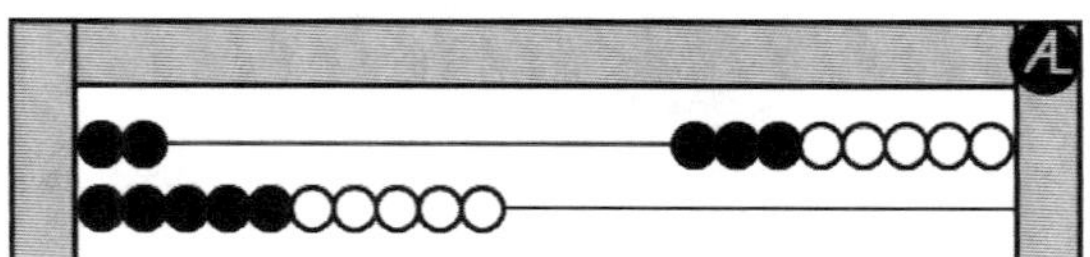

FUTURE REVIEW. Write the facts and ask various children to recite the sums. If they are having a problem with, for example, 8 + 9, ask, **What is 8 + 10?**

An alternate way to help the children to learn these facts is to turn the abacus sideways. Write

$$\begin{array}{r} 9 \\ +\ 3 \\ \hline \end{array}$$

3	+	9	=	
7	+	9	=	
9	+	9	=	
9	+	2	=	
6	+	9	=	
9	+	8	=	
9	+	5	=	
2	+	9	=	
9	+	6	=	

Demonstrate entering 9 and 3 as shown and moving up a bead to form 10. Repeat with other examples. See the corresponding worksheet (4-26).

ORAL PROBLEMS. A. Kelsey walked 9 blocks in the morning and 6 blocks in the afternoon. How many blocks did she walk? [15 blocks]

B. Mrs. Clark's class had a hamster. The hamster had 4 babies in the spring and 9 babies in the fall. How many babies did she have that year? [13 babies]

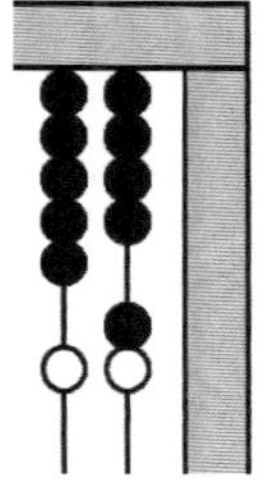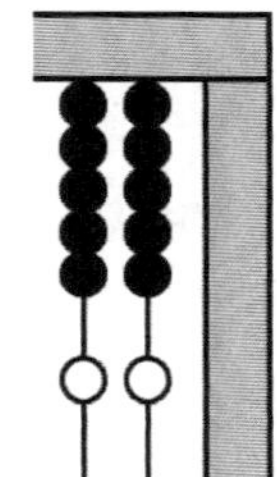

2. Write on the chalkboard 36 + 9. Ask the children to enter 36 with the units separated; that is, the units are entered on the last two wires as shown. Ask them to perform the Take and Give procedure to make another 10 and finally to move that 10 to the other 10s by another Take and Give. Call on a child to record the answer. Repeat with 53 + 9 and 49 + 4.

	9		9		9		9
+	3	+	2	+	4	+	6

	9		9		9		9
+	9	+	1	+	7	+	5

	9		9		9		9
+	8	+	0	+	3	+	5

ORAL PRACTICE. Write and say for the children the following:

$$\begin{array}{cccccc} 48 & 29 & 16 & 23 & 69 & 77 \\ +9 & +5 & +9 & +9 & +4 & +9 \end{array}$$

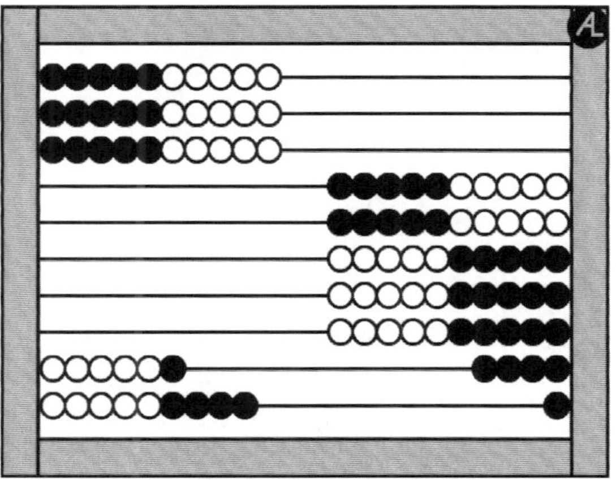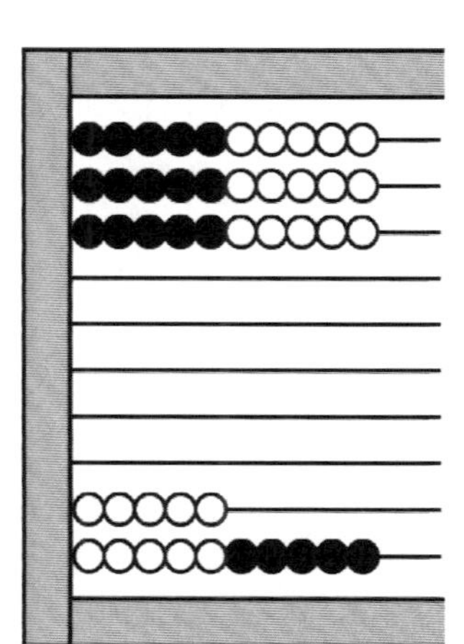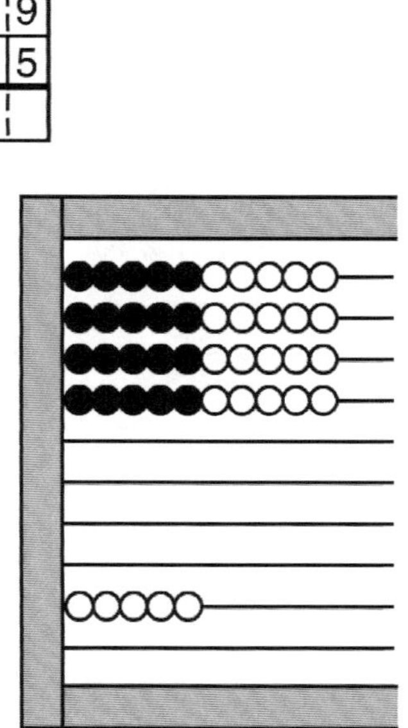

When these are added mentally, first think of adding 10 and then going back 1. Give them worksheets similar to these problems.

The worksheet (4-27) at the right asks for adding two-digit numbers.

Adding 8 to a number

To add 8 to an even number involves thinking of the next lower even number (subtracting 2 or counting back 2) and then adding 10. Likewise, to add 8 to an odd number, think of the next lower odd number and then add 10.

Start by reviewing the evens in reverse order: 10, 8, 6, 4, 2, 0. Do this with the abacus; enter 10 and remove 2 at a time while stating the remainder. Also review the odds in reverse order: 11, 9, 7, 5, 3, 1.

Write on the board

$$\begin{array}{cccccc} 8 & 8 & 7 & 8 & 4 & 8 \\ +5 & +3 & +8 & +6 & +8 & +8 \end{array}$$

and ask the children to enter 8 and 5 on the first two wires of their abacuses. Ask them to use Take and Give to find the answers. Repeat for the other problems.

Give similar problems for oral practice and for worksheets (4-28).

As an alternate visual model, use the abacus with the wires vertical and give them the worksheet (4-29 and 4-30) with column addition.

Write on the chalkboard, 46 + 8. Ask the children to enter 46 with the units separated and then to add the 8. Write more problems for the children to do.

$$\begin{array}{cccccc} 46 & 39 & 57 & 29 & 14 & 87 \\ +8 & +5 & +9 & +6 & +9 & +9 \end{array}$$

Give them worksheets similar to these. The worksheet (4-31 and 4-32) on the right includes facts for + 9 as well as + 8.

ORAL PROBLEMS. A. Marietta had 38 leaves in her collection. She found 7 more. How large is her collection now? [45 leaves]

B. Timmy spends 8 minutes walking to the bus and 19 minutes on the bus. How long does it take him to get to school? [27 minutes]

C. The hall at West School had 48 pictures on the wall. Nine more were put up. How many pictures are up now? [57 pictures]

36	+	9	=	
58	+	9	=	
49	+	3	=	
61	+	9	=	
19	+	6	=	
29	+	5	=	
14	+	9	=	
89	+	4	=	
63	+	9	=	

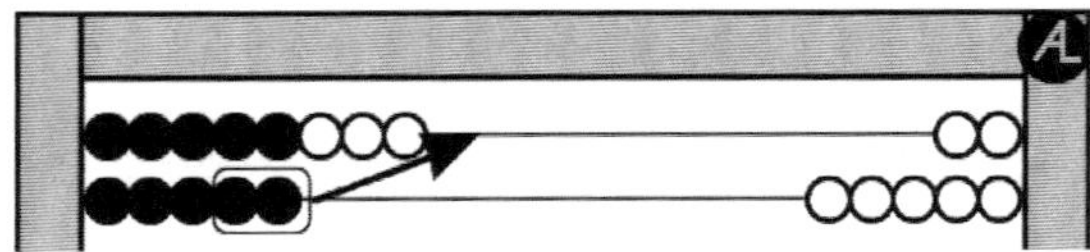

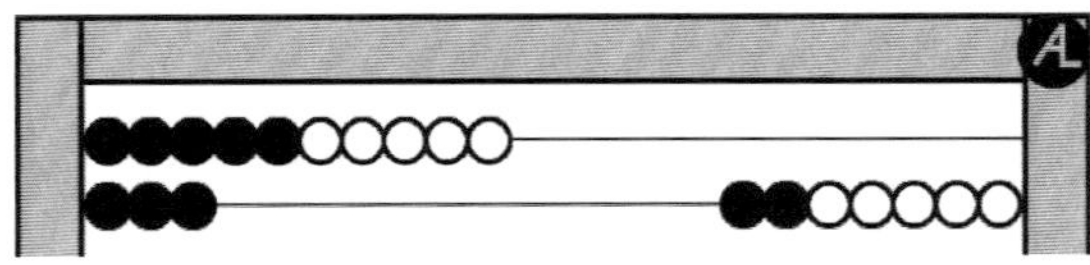

8	+	5	=	
8	+	6	=	
3	+	8	=	
8	+	4	=	
7	+	8	=	
8	+	8	=	
8	+	9	=	
2	+	8	=	
1	+	8	=	

$$\begin{array}{cccc} 8 & 8 & 8 & 8 \\ +6 & +3 & +9 & +5 \end{array}$$

$$\begin{array}{cccc} 8 & 8 & 8 & 8 \\ +8 & +0 & +1 & +7 \end{array}$$

$$\begin{array}{cccc} 8 & 8 & 8 & 8 \\ +2 & +4 & +3 & +5 \end{array}$$

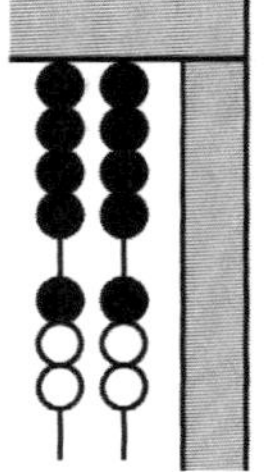 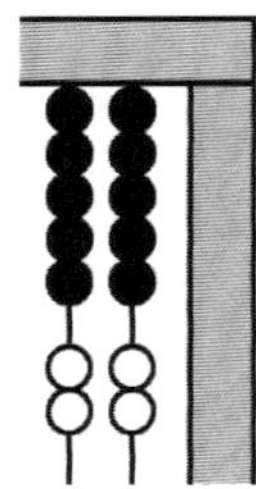

57	+	8	=	
18	+	9	=	
21	+	8	=	
34	+	8	=	
88	+	2	=	
48	+	3	=	
56	+	8	=	
68	+	5	=	
78	+	4	=	

58	+	3	=	
64	+	9	=	
26	+	9	=	
39	+	2	=	
78	+	9	=	
47	+	8	=	
56	+	8	=	
68	+	6	=	
28	+	3	=	

Unit 5
Introducing subtraction

Although it is not necessary for the children to have mastered all the addition facts before beginning subtraction, they will learn subtraction much more easily if they thoroughly understand addition. This unit introduces subtraction; the children are not expected to learn the facts at this point, although they will learn some of them. The strategies for mastering the facts are reserved for the next unit. An algorithm for paper and pencil subtraction will be taught still later. It is different from the standard one found in the textbooks in the United States and is simpler to learn. Work proceeds from left to right, as it does in division.

Addition was taught as increasing a given quantity; now subtraction will be taught as decreasing the given quantity. The process of finding the missing addend, introduced in an earlier unit, while written as an addition equation, requires subtraction for solution. Subtraction also is the process needed to compare two quantities, or find their difference.

Some new terms are necessary to discuss subtraction. In the example, $8 - 6 = 2$, 8 is the minuend, 6 is the subtrahend, and 2 is either the remainder or the difference. Whether the answer is a remainder or a difference depends upon the nature of the problem. Obviously the children will not need these terms to start.

At this point, some texts introduce the children to the term "number sentences." The term "sentence" or "number sentence" for equation is unfortunate. The word never has this meaning outside the primary classroom. Also, children of this age have little comprehension of a grammatical sentence, which makes the analogy inappropriate. Use of the term equation both enhances mathematical vocabulary and avoids problems in the language class.

Another unfortunate term is "take away," which implies going back or going down. Sometimes, subtraction is accomplished by going up. If someone buys an article costing 37¢ with 50¢, the change is figured by starting at 37, going up to 40, and then going up to 50. Likewise, to find $41 - 39$, it is much simpler to start at 39 and go up to 41, while observing the amount to go up was 2. On a grammatically note, in the phrase "6 take away 4," is 6 taking 4 away?

Although most textbooks today ignore the word "borrow," it has several advantages over its substitutes. First, it is reserved exclusively for subtraction; whereas, *trading, regrouping,* and so forth, are also used for addition. Secondly, *borrow* aptly describes the process much as one borrows an idea. Thirdly, the dictionary recognizes this meaning of *borrow,* but not any of the substitutes. It is not in any danger of extinction since it is part of the vocabulary of computer programmers.

SIMPLE SUBTRACTION

In this first practice all the quantities are less than 10.
Mastery of the facts is not expected at this point.

Oral subtraction

1. Approach the idea of subtraction as the opposite of addition. Ask, **What is the opposite of hot? [cold] What is the opposite of big? [small] What is the opposite of small? [big] Well, the opposite of add is subtract. What is the opposite of add?** [subtract] **What is the opposite of subtract?** [add]

Prepare a stack of books or blocks that the children can see. Ask a child to place a few more books on top and ask the group, **What is he doing when he puts more on top?** [adding] Ask another child to remove some books and say, **What is she doing when she removes some?** [subtracting] Repeat the adding and removing several more times and ask the children to name the operation.

2. To help the children with the terminology "subtract from," arrange the following scenario. Give each of several children at least six items. Then ask a child to subtract 2 blocks from Angie's 6 blocks or 3 cars from Jamie's 5 cars. Summarize by saying that the first thing they had to do was to go to Angie's blocks or Jamie's cars and then they could do the subtracting.

Tell the children that they are going to start subtraction on their abacuses. **Instead of making a number bigger by adding to it, we are going to make it smaller by subtracting from it.**

Enter 6 on your abacus and say, **This is how to subtract.** Subtract 2 by sliding the beads to the right as shown and say, **We have subtracted 2 from 6,** or **We have started with 6 and subtracted 2.** Ask them to subtract 3 from 6. If necessary, coach them by asking, **Which number do we enter?** [6] Repeat with other oral examples such as 7 subtracted from 8 and 1 subtracted from 5.

Written subtraction

Ask the children if they are wondering how we write a subtraction problem. Write the sign "−" and tell them that it is the minus sign.

Write 8 − 5 = and say, **We read this as eight minus five equals. What is the opposite of minus?** [plus] **What is the opposite of plus?** [minus] **Use your abacus and find 8 − 5.** [3] Write more examples, such as 9 − 2, 5 − 3, 7 − 1, 10 − 5, and 6 − 4. Invite individual children to write the answers on the board.

Give them worksheets (5-1 and 5-2) to do over a period of several days. To emphasize the new topic of subtraction, the worksheets will no longer have little rectangles for each element in the equation, only a larger rectangle for the entire equation.

Subtracting from 10

1. Tell the children to enter nine 10s. Ask them to subtract 1 from the first row; call on a child to recite the fact and another child to write it. [10 − 1 = 9] Ask the children to subtract 2 from the second row; call on other children to state and write the fact. [10 − 2 = 8] Continue to 10 − 9.

Give them blank sheets (5-3A) to repeat the activity on their own. Also give worksheets (5-3B) with random facts.

6	−	2	=
4	−	1	=
9	−	7	=
8	−	5	=
5	−	4	=
4	−	2	=
7	−	1	=
10	−	2	=
7	−	4	=

10	−		=
10	−		=
10	−		=
10	−		=
10	−		=
10	−		=
10	−		=
10	−		=
10	−		=

10	−	8	=
10	−	4	=
10	−	3	=
10	−	1	=
10	−	7	=
10	−	6	=
10	−	2	=
10	−	5	=
10	−	9	=

2. Do the same activity using the abacus with the wires vertical.

After they have completed the worksheets (5-4), ask whether or not they expected the answers they found. You might ask, **Two and what equals 10? And 10 – 2 is what?** [8] Repeat with 7 + ? and 10 – 7 and with 4 + ? and 10 – 4 until they understand the similarity.

ORAL PRACTICE. Write the facts on the board and have the class perform "choral" practice or ask them to practice with partners.

Subtracting 1

1. Review by asking the children to count backwards from 10 to 1. Then write or say a number and ask them to say as quickly as they can the number that comes before it. If more help is needed, have them play the game, Guess What's Before, a variation of Guess What's Next described in Unit 1.

Ask the children to build the stairs. For each wire, ask the children to subtract 1 and call on a child to recite the fact. Alternately, the facts can be constructed from a single 10 by successively subtracting 1 as shown. Then ask them, **What is the rule for subtracting 1 from a number?** [The answer is one less than the original number.] Give them a worksheet (5-5) for practice.

This is also a good time to help the children become alert for watching the minus sign. It is amazing to see the number of errors junior high students make in computation by performing the wrong operation. Therefore, also give them a worksheet with both adding and subtracting. You might want to practice on the board with both types of problems.

2. The children also need to work on subtracting 1 in the vertical format. But this time ask them to use the first two wires. Later on, that will be the position for thousands, so it is time for them to practice entering and reading quantities in this position. The previously used position of the two right wires will be the 1s place and it needed to be learned first to seem the most natural.

Ask them to enter 10 on the left two wires. Ask the children to subtract 1 and recite the fact. Continue to 1 – 1. Provide worksheets (5-6 and 5-7) with only subtraction and with both addition and subtraction.

ORAL PROBLEM. Melissa was carrying 6 colored pencils. She dropped one. How many does she still have? [5 pencils]

First the children place the 6 and 1 in the correct places on the part/whole circles. Next they solve the problem on the abacus and fill in the remaining circle. Lastly, they write the equation. Learning to write a solution in an understandable form is an important part of problem solving.

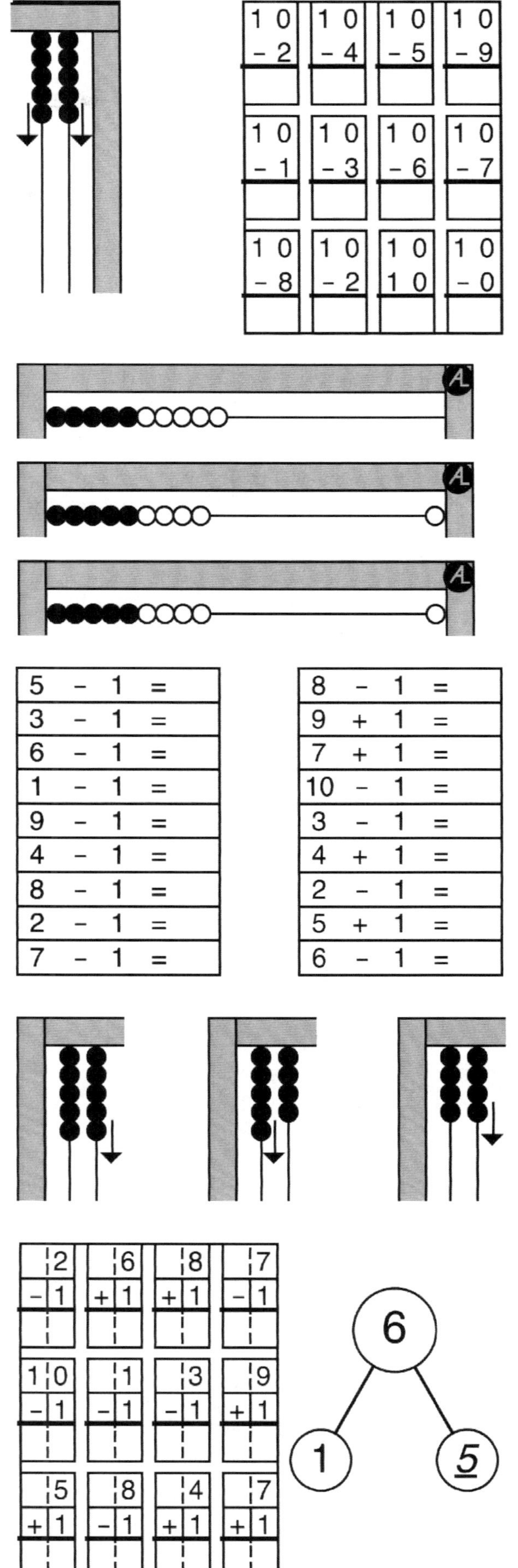

SUBTRACTIONS > TEN

There is no reason to delay using numbers beyond 10 for subtracting. The use of the higher decades provides reinforcement as well as more examples for subtraction.

Subtracting tens and ones

1. Review that 10 tens is called one hundred and is written 100.

Write on the board

$$60 - 20 =$$

and ask the children how they would subtract 20 from 60. If they need help, ask them first to enter 60. The 20 is subtracted by sliding the lower 2 tens back to the right. You might want to stop a short ways from the right edge to indicate the original problem, but the children usually want to slide them all the way to the edge. Ask them to solve 100 – 40 and 80 – 60.

Give them a worksheet (5-8) with similar problems. From now on, throw in an occasional addition problem to keep them alert to watching the signs.

ORAL PROBLEM. A. Karl had 50 cards. He gave 10 cards to Jackie. How many does he have left? [40 cards]

B. Jackie had 30 cards. She received 10 cards from Karl. How many does she have now? [40 cards]

2. The children can also perform subtractions with problems such as 64 – 2 and 64 – 20.

Write for the children

$$57 - 4 =$$

Tell them to enter the 57 this time with the ones separated, that is, on the bottom row. To subtract the 4, it is removed from the 7 and the answer read as 53. Repeat with 48 – 3.

Give them another type of example

$$49 - 20 =$$

Again the ones are separated, but this time the 2 tens are subtracted from the 4 tens. The answer is read as 29.

Write similar problems on the board for the children to work with partners. Or give them worksheets (5-9) to work together.

ORAL PROBLEM. A. Cyndi had 55 cents. If she spent 23 cents, how much does she have left? [32 cents]

B. Jordan had 86 cents and spent 52 cents. How much does he have left? [34 cents]

C. Chelsey is reading a book with 67 pages. If she has read 43 pages, how many does she have left? [24 pages]

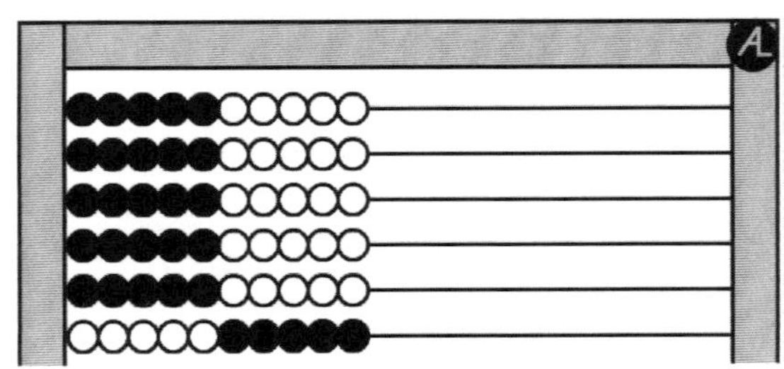

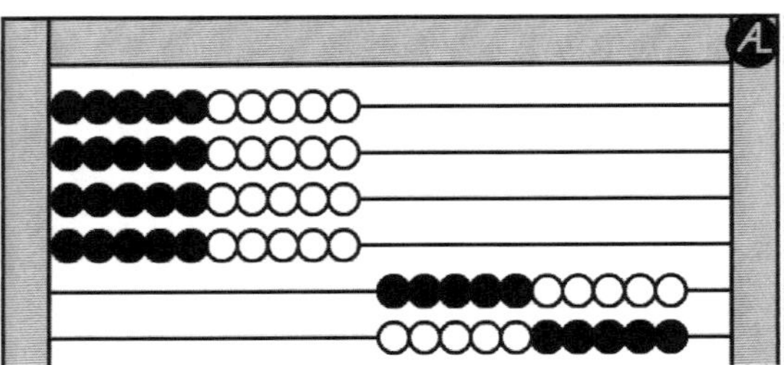

60	– 20	=
50	– 30	=
90	– 40	=
80	– 10	=
10	– 0	=
70	– 60	=
30	+ 20	=
70	– 50	=
60	– 10	=

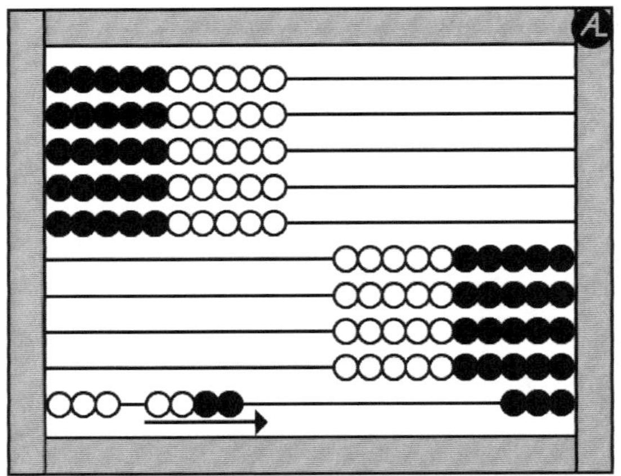

57	– 4	=
76	– 3	=
95	– 30	=
33	– 10	=
42	+ 1	=
38	– 20	=
63	– 30	=
77	– 2	=
36	– 5	=

Subtracting single-digit numbers

Subtracting 6 from 14 presents the dilemma of subtraction: borrowing, also known ,of course, as *trading, exchanging, regrouping, renaming,* and other terms. The only terms that are likely to be within the experience of primary children are *trading* and *borrowing*.

1. Write on the board

$$14 - 6 =$$

and ask the children to enter 14. The 6 is subtracted by first removing the 4 beads from the second wire and then removing the last 2 from the right in the top row. Repeat for $11 - 2$ and $16 - 9$.

It is probably more instructive to perform these subtractions with the wires vertical. Write on the board

$$13 - 6 =$$

and ask the children to enter 13 with the wires vertical and then to subtract 6. Repeat for $16 - 8$ and $18 - 9$.

Give them worksheets (5-10 and 5-11) in each format.

ORAL PROBLEMS. A. Nathan's class has 13 boys. Six of the boys wore jackets to school one day. How many boys did not wear jackets? [7 boys]

B. Estee has 15 cousins. Nine of her cousins are boys. How many are girls? [6 cousins]

C. Joyce worked 4 hours on Monday, 6 hours on Tuesday, and 5 hours on Thursday. How many hours did she work? [15 hours]

D. Mr. Smith's class has 22 students. One day 2 were absent. How many were there? [20 students]

2. Another set of examples which provides an excellent introduction to borrowing is problems of the type $50 - 3$ or $80 - 4$.

Write on the board

$$50 - 3 =$$

and ask the children to solve the problem. Here they cannot subtract tens from tens nor ones from ones, but the ones have to be subtracted from one of the tens. After a few more examples such as $80 - 4$ and $100 - 5$, they can work on a worksheet (5-12) independently or in groups.

ORAL PROBLEMS. A. Kyle counted 30 sheets of paper for an art project. Four of the sheets were torn and could not be used. How many good sheets were there? [26 sheets] If there were 24 children in the class, was there enough paper? [yes]

B. The children at Highland Elementary had 20 minutes for recess. If it took them 4 minutes to get ready, how many minutes did they have to play? [16 minutes]

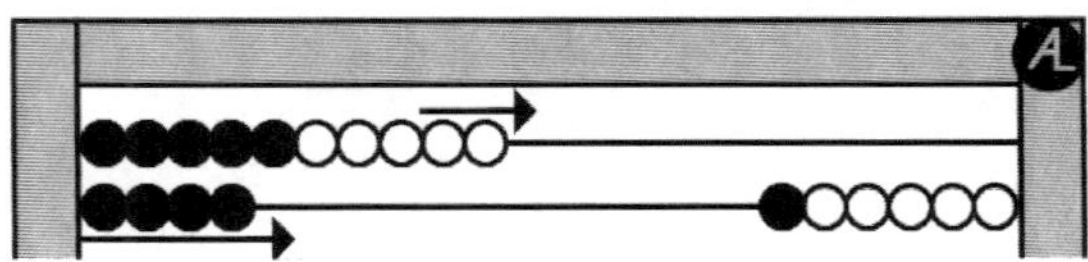

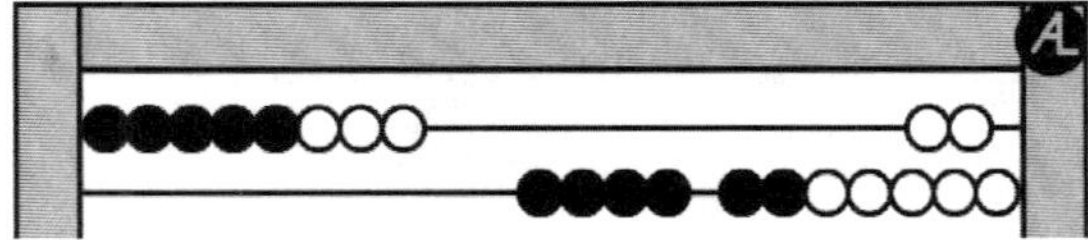

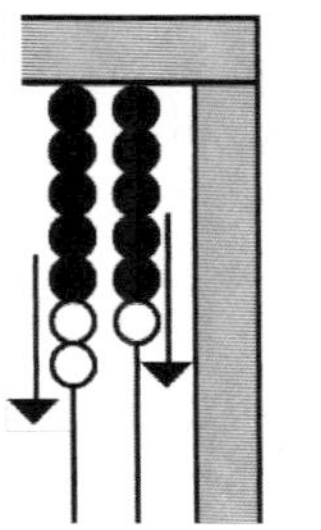
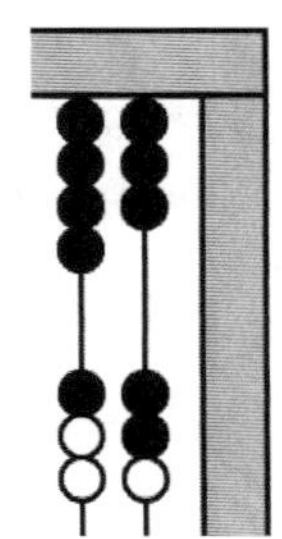
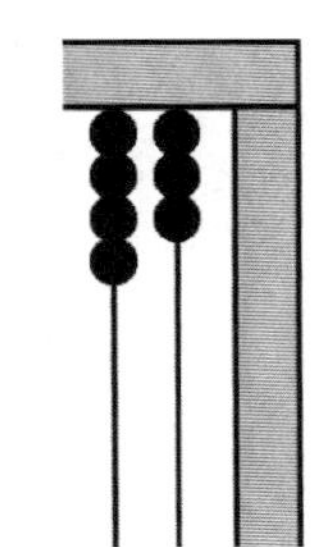

14 – 6 =			
13 – 8 =			
11 – 2 =			
16 – 9 =			
15 – 6 =			
12 – 5 =			
11 + 7 =			
12 – 6 =			
13 – 8 =			

13	11	14	16
– 6	– 9	– 5	– 8
17	12	11	13
– 9	– 6	– 7	– 8
12	11	14	12
– 4	– 3	– 8	– 5

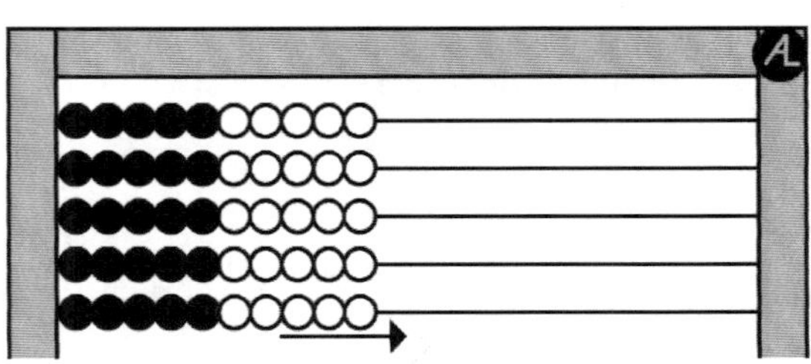

50	– 3	=	
40	– 9	=	
30	– 7	=	
90	– 1	=	
10	– 6	=	
20	– 8	=	
80	– 2	=	
70	– 5	=	
30	– 9	=	

3. Teach the final case of subtracting a single-digit number from a number between 20 and 99. Write on the board

$$43 - 6 =$$

Tell the children to enter the 4 tens as usual but to enter the 6 ones on the bottom wire. Then say, **Since it is impossible to subtract 6 from the 3, we will take a ten from the tens and give it to the ones. So how many ones will we have?** [13] **Now we can subtract 6. What is the answer?** [37] **What happened to the number of tens?** [It is 1 less.] The steps are shown in the figures on the right.

Repeat for 25 – 7 and 62 – 4. This could be an activity for two; one person does the work for the tens, including writing the answer, and the other does the work for the ones. A sample worksheet (5-13) is shown.

ORAL PROBLEMS. A. On New Year's day, Jim counted 57 days until his birthday. Nine days have gone by. How many more days must he wait? [48 days]

B. Anna planted 24 tulips. Early in the spring, 9 could be seen growing. How many more should she soon expect to see? [15 tulips]

Subtracting double-digit numbers

1. The first case with the subtrahends less than the minuends presents no difficulty. The tens are subtracted from the tens and the ones from the ones.

It is neither necessary nor desirable for children to think that subtraction must always start at the right. Calculators, computers, and frequently mental work all start from the left. In a later chapter, a simpler method of subtraction is given where starting with the left is the norm.

Present the children with the problem

$$76 - 23 =$$

Ask them how to they would do it. If necessary, remind them to enter the ones on the bottom wire. The figures show the sequence: the 2 tens are subtracted from the 7 tens and the 3 ones are subtracted from the 6 ones.

Give the children other examples to try, such as, 65 – 24 and 44 – 41. In the latter example, tell them that we do not write (14 and 5-15).

ORAL PROBLEM. A. Jennie climbed up 53 steps. She climbed back down 27 steps. How many does she have to climb before reaching the bottom? [26]

B. Benji rode with his family to visit his grandparents, who lived 56 miles away. When they had traveled 32 miles, they stopped for fuel. How far did they still have to go? [24 miles]

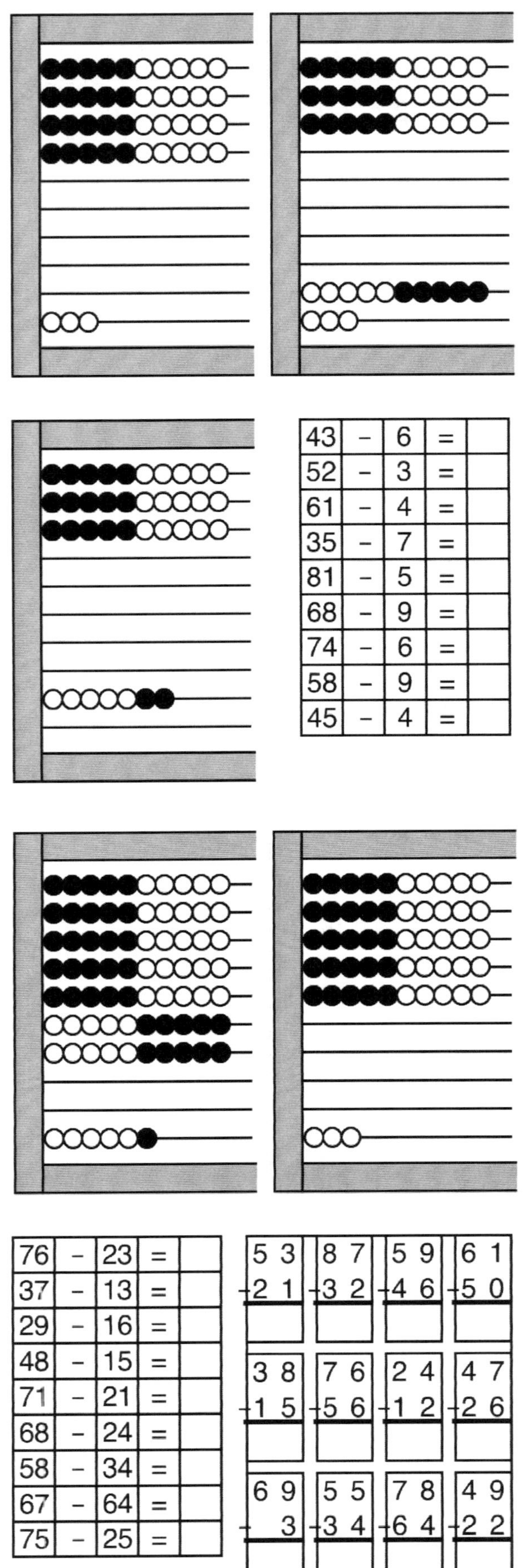

43	–	6	=	
52	–	3	=	
61	–	4	=	
35	–	7	=	
81	–	5	=	
68	–	9	=	
74	–	6	=	
58	–	9	=	
45	–	4	=	

76	–	23	=	
37	–	13	=	
29	–	16	=	
48	–	15	=	
71	–	21	=	
68	–	24	=	
58	–	34	=	
67	–	64	=	
75	–	25	=	

5 3	8 7	5 9	6 1
–2 1	–3 2	–4 6	–5 0

3 8	7 6	2 4	4 7
–1 5	–5 6	–1 2	–2 6

6 9	5 5	7 8	4 9
– 3	–3 4	–6 4	–2 2

2. This final case combines all the previously learned subtraction techniques. Write on the board the following:

$$62 - 34 =$$
$$39 - 16 =$$
$$70 - 53 =$$
$$45 - 28 =$$

and ask, **Which problem do you already know how to solve?** [39 – 16]

To solve the first problem, ask the children to enter 62 with the tens and ones separated. Ask, **Can you subtract the 3 tens?** [yes] **Can you subtract the 4 ones?** Some of the children might think of bringing a ten down to the ones. The four steps are shown in the accompanying figures.

Ask the children to solve the three remaining problems. In the third problem 70 – 53, it is not necessary to bring a ten down before subtracting the 3.

Give the children worksheets (5-16) for practice. Include sheets with the problems written vertically, which are to be solved the same way.

ORAL PROBLEM. Miguel is reading a book with 56 pages. He has read 38 pages. How many pages does he have yet to read? [18 pages]

SUBTRACTION RESULTS

Addition answers only one question: how many. Subtraction, on the other hand, answers three questions: what's left, the difference, and how much more.

Greater than, equal to, or less than

Children understand these concepts at a very young age, but may have difficulty with the words and especially the > and < symbols. In some cases, confusion over reading the symbols persists into the junior high years and beyond. Therefore, wait until the left to right sequence is thoroughly established and reversals have disappeared before teaching how to read the symbols.

1. Introduce the term, GREATER, with two stacks of blocks by asking, **Which stack is greater?** Explain that greater means more. Use other concrete objects for comparison before writing numbers.

To teach the relationship between the terms, greater than and less than, compare to older and younger. Ask, **Who is older, you or I? Who is younger?** Use other familiar examples until the children understand the nature of opposite. Include a case of twins to introduce the "equal" condition.

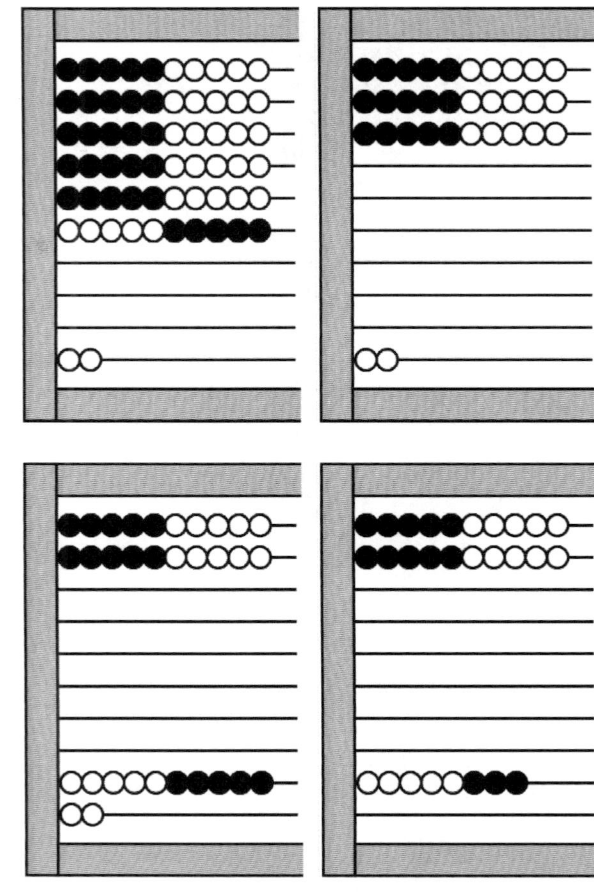

62	–	34	=	
76	–	28	=	
80	–	53	=	
61	–	48	=	
26	–	17	=	
48	–	25	=	
71	–	52	=	
45	–	19	=	
65	–	38	=	

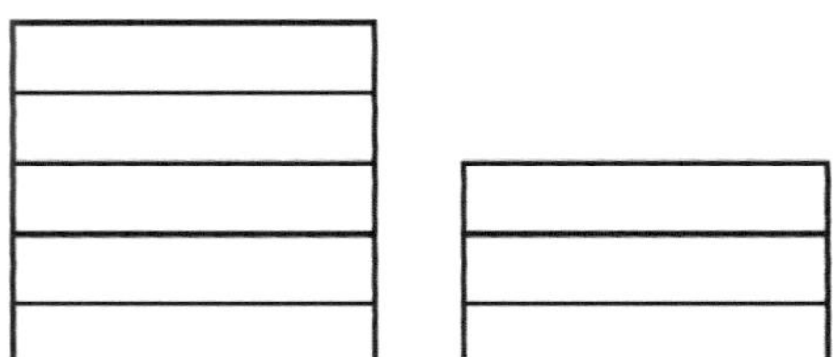

What is the opposite of older? [younger] **Well,
the opposite of greater than is less than? What
is the opposite of greater than?** [less than] **What is
the opposite of less than?** [greater than] Refer to the
stacks of blocks: **This stack is greater than that
stack, and that stack is less than this stack.**

Ask the children to enter 4 on the top wire of their abacuses. Then ask them to enter a quantity more than 4 on the
second wire. Discuss the possible answers. [5 through
10] With 4 on the top wire ask them to enter a number
less than 4. Again discuss the possible answers. [1 to 3]
Finally, ask them to enter a number equal to 4. Ask,
How many answers are there this time? [only one,
4]

2. Write on the board

        8 5   3 4   7 7

and tell them that today they will learn how to write the
symbols that show comparing each of the three groups:
Which one needs the equal sign? Call on a child to
write the "=" sign between the 7s.

Use this simple way to teach writing the signs that virtually all children can follow. Ask them to draw two dots at
the top and bottom of the greater number and one dot at
the middle of the lesser number. Then tell them to connect
the dots starting at the top.

Later on the dots can be eliminated, but the symbol is
drawn by starting at the greater number, proceeding to the
lesser number, and ending at the greater number. Write
several pairs of numbers with circles between them and let
the children write the correct sign.

Give them the left worksheet (5-17). Later, include numbers up to 99 (5-18).

To help the children read the symbols, draw them on the
board and cover all but a small portion of the left edge.
Ask, **Which one shows a large amount?** [>] **So we
read it as "is greater than."** Then ask, **Which one
shows a small amount?** [<] **So we read it as "is
less than."** Write several of the symbols. Point to the
symbols and have the children practice reading them.

Later ask them to work in pairs to do the sheet (5-19) on
the right where reading the symbols is necessary. The
problems can be read as "10 is greater than what number?" and "What number is greater than 9?" Many different answers are possible.

Remainders and differences

1. Write on the board

$$9 - 7 =$$

Tell the children, **To solve this problem we start
with 9 and then subtract 7. What remains?**

3	O	2
7	O	8
3	O	5
9	O	2
1	O	4
7	O	7
10	O	1
2	O	6
6	O	0

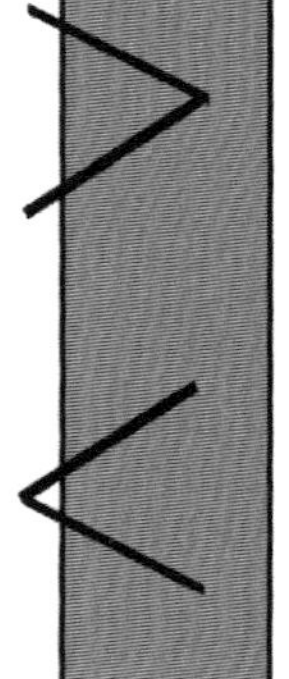

10	>	
6	<	
3	=	
8	<	
	>	9
	<	4
	<	2
	=	5
	>	0

"Remains" means what is left. [2] **The 2 is called the remainder.**

Write several examples on the board and ask the to identify the remainders:

$$8 - 5 = 3 \qquad 10 - 4 = 6 \qquad 9 - 2 = 7$$

2. Ask the children to enter 8 and 6 on the top two wires. Then ask, **How can we find the difference between the number of beads on the two rows?** Explain that one way is to cover up what they have in common, or what they both have, which leaves 2.

For numbers over ten, turn the abacus sideways and use two sets of adjacent wires as shown. Write the numbers 9 and 14 and tell the children to find the difference. Show them how to enter 9 and 14. Ask, **How many beads will have to be covered.** [9] One of the dark beads from the 14 must be moved down to avoid being covered with the 9. Give them more examples, such as finding the difference between 16 and 12 or between 11 and 6.

3. After the children have completed the worksheet (5-20), write 7 and 10 and ask them to find the difference. Ask them what operation helps them to find it. [subtraction] After they understand that subtraction is the operation, they are ready for oral problems and their written solution. Remind them that in writing a problem, the larger number must be written first.

ORAL PROBLEMS. A. Hang can name 11 capitals and Vang can name 8 capitals. What is the difference in the number of capitals they can name? [3 capitals]

B. Kara's dog is 13 inches high and Patrick's dog is 17 inches. What is the difference in height between the two dogs? [4 inches]

C. Ms Green's class has 34 reading books while Ms Black's class has 42 books. What is the difference in the number of books? [8 books]

D. The students in Waterville have 92 days of summer vacation, but the students in Huntstown have 86 days. What is the difference in the amount of vacation? [6 days]

Subtraction twins

For want of a better term, "subtraction twins" is used here. However, we rarely have need to discuss the concept once it is understood.
Write on the board

$$8 - 3 =$$
$$8 - 5 =$$

Demonstrate first to the children this slightly different way of subtracting. Enter 8 and slide it a little ways away from the left edge, as shown. To show $8 - 3$, use either your hand or a card to cover the 3. To show $8 - 5$, cover the 5. When working alone, the children may prefer to

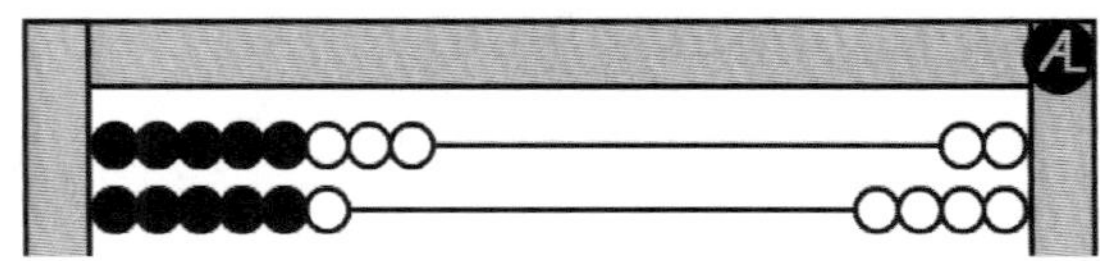

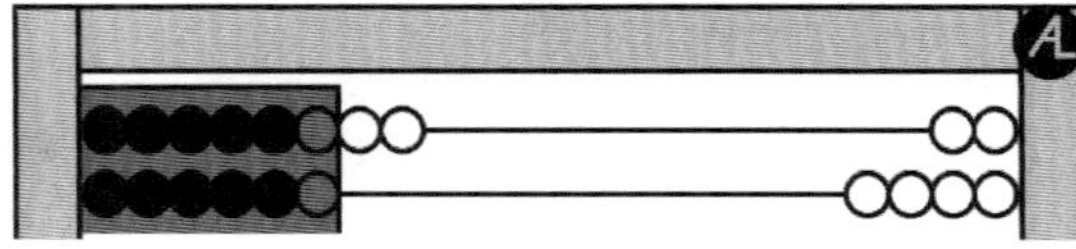

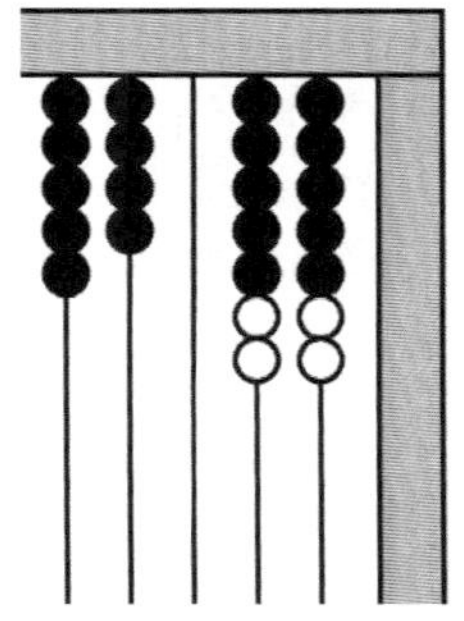
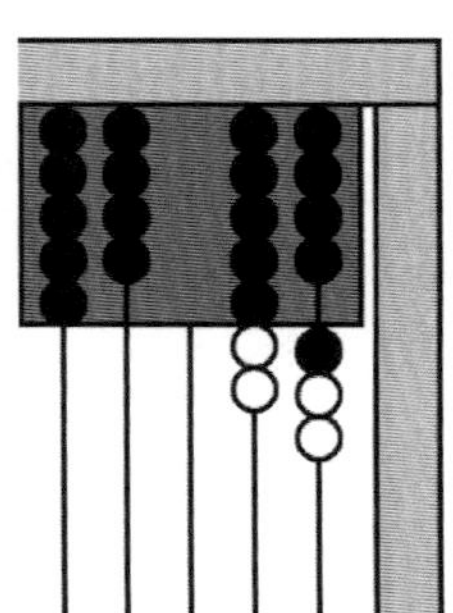

Find the difference.		
6 and	8	___
9 and	14	___
10 and	7	___
5 and	9	___
15 and	10	___
19 and	8	___
12 and	11	___
4 and	8	___
15 and	8	___

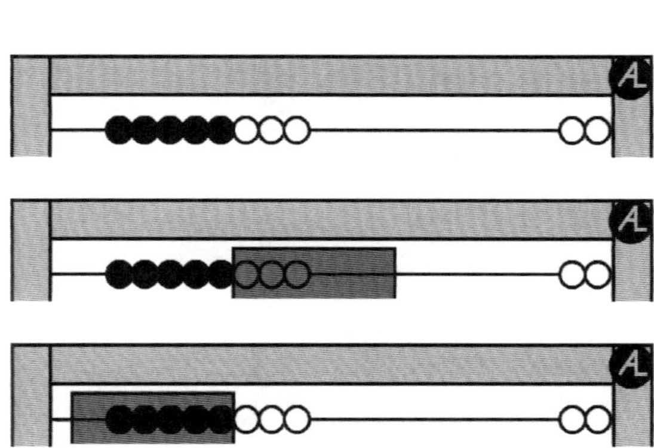

slide the quantities to the respective sides of the abacus.

Next, show them how to do the activity with 11 – 8 and 11 – 3. See the accompanying figures. Give the children the following combinations: 6 – 1 and 6 – 5 and also 10 – 4 and 10 – 6.

Give the children a worksheet (5-21) for practice with the abacus. Also give them a worksheet (5-22 and 5-23) in both formats where they find and write the other subtraction twin.

ORAL PROBLEMS. A. Ramon is affixing stamps to 25 greeting cards. If he has done 18 cards, how many remain to be done? [7 cards]

B. Jessica is weeding the garden with 14 rows. If she has weeded 9 rows, how many are left? [5 rows]

C. Carlos counted 11 busses traveling north and 9 traveling south. How many buses did he see? [20 buses]

D. Marietta wanted to read 40 books over the summer. By July 4 she had read 29. How many more did she need to read? [11 books]

E. Naruib sold 10 apples on Monday, 19 apples on Wednesday, and 10 apples on Friday. How many did he sell for the week? [39 apples]

F. Salim rode his bicycle 13 blocks. He rode 5 blocks after a rest. How many blocks did he ride before his rest? [8 blocks]

Equations

Tell the children that today will be a puzzle day. Write on the board

$$3 + 2 = 5 \qquad 10 - 6 = 4 \qquad 3 + 5 + 7 = 15$$

and tell them that these are equations.

Next write on the board

$$9 + 5 \qquad 17 \qquad \begin{array}{r} 10 \\ -4 \end{array}$$

and tell them that these are not equations. Now ask them what an EQUATION must have. [an equal sign.]

Tell them, **Here is your first puzzle. I will write three numbers and you are to think of at least two equations using these numbers.**

Write 3, 5, and 8. Call on one or two of the children to write the equations. [3 + 5 = 8 and 5 + 3 = 8]

If they have not included 8 – 3 = 5 and 8 – 5 = 3 as equations, give them the numbers 8, 5, and 3 and ask for two more equations. Ask the children to show all four prob-

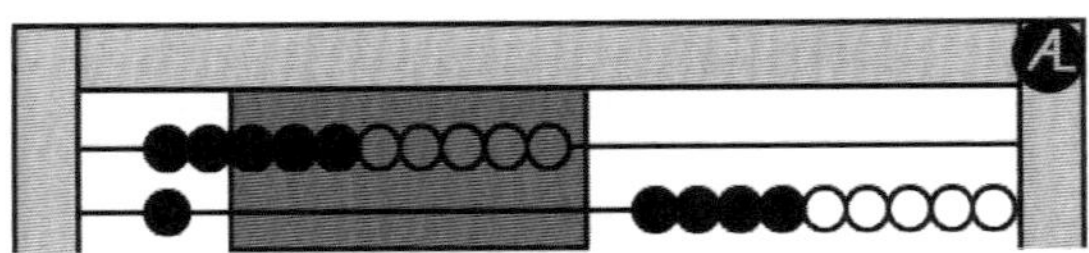
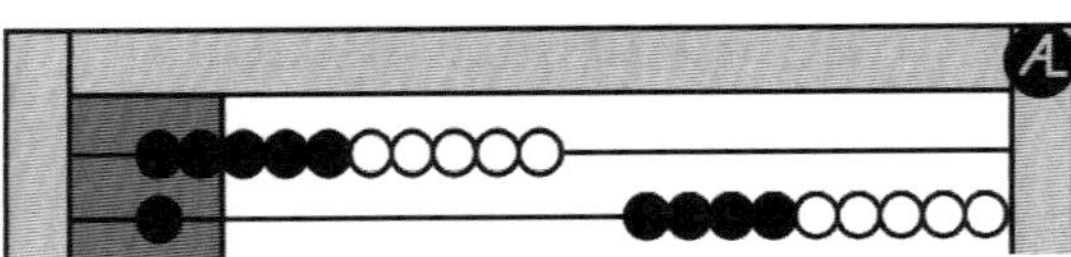

8 – 3 =	7 – 5 =
8 – 5 =	
13 – 6 =	8 – 1 =
13 – 7 =	
6 – 4 =	10 – 4 =
6 – 2 =	
9 – 6 =	12 – 8 =
9 – 3 =	

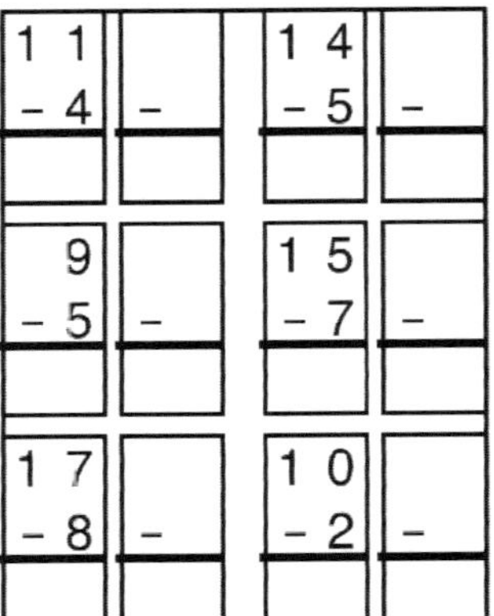

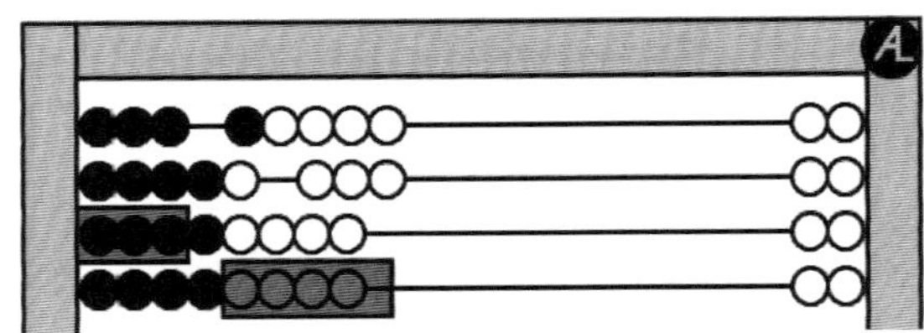

Write equations using 3, 6, 9.

+	=
+	=
–	=
–	=

Write equations using 2, 8, 10.

+	=
+	=
–	=
–	=

lems on the abacus; this helps them realize concretely the relationship between addition and subtraction.

Give them more puzzles, asking them to find four equations for each set of numbers: 5, 6, 11 and 6, 7, 13.

As a teaser, ask for all the equations for 5, 5, 10. [Only two exist.]

A sample worksheet (5-24 and 5-25) is shown.

Checking

The concept that subtraction can be checked by adding the remainder to the subtrahend is difficult for many children.

Write on the board

$$\begin{array}{r} 15 \\ -7 \\ \hline \end{array}$$

and ask the children to perform the subtraction with the wires vertical. Call on a child to write the remainder. Then ask what would happen if the remainder, 7, were added back to the 8. [It must equal what we started with, 15.]

Let them try a few more combinations such as

$$\begin{array}{r} 9 \\ -3 \\ \hline 6 \end{array} \qquad \begin{array}{r} 11 \\ -4 \\ \hline 7 \end{array}$$

before proceeding.

Write on the board the following problems

$$\begin{array}{r} 10 \\ -2 \\ \hline 8 \end{array} \qquad \begin{array}{r} 8 \\ -4 \\ \hline 3 \end{array} \qquad \begin{array}{r} 17 \\ -9 \\ \hline 9 \end{array} \qquad \begin{array}{r} 34 \\ -18 \\ \hline 16 \end{array}$$

Start with the first problem. Cover the top number with a card and ask the children to add the remaining two numbers. Follow the sequence at the right. Next remove the card and ask the children to compare the top and bottom numbers, **Are they the same?** [yes] **That means that the remainder 8 is correct.**

Do the same for the next problem. However, here the top and bottom numbers do not agree. Ask the children to do the subtraction and correct the mistake.

Do the next two problems with the children. Let them decide what method of adding they wish to use. Then provide them with worksheets (5-26) that ask them to find the subtraction mistakes.

Encourage them in the future to check their own work or have them work with partners and check each other's work. In some cases, the addition can more easily be performed on another piece of paper. Or, after a row of subtraction is completed, the paper can be folded to hide the minuends. The whole row is then added and compared.

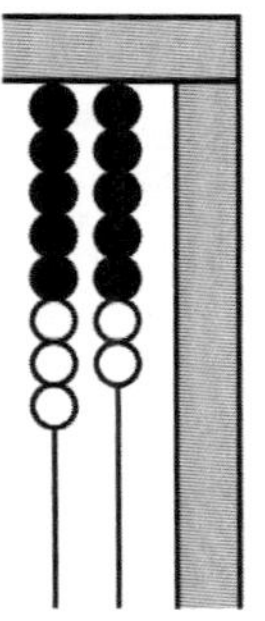
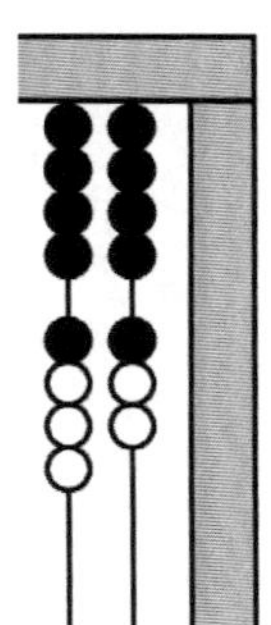
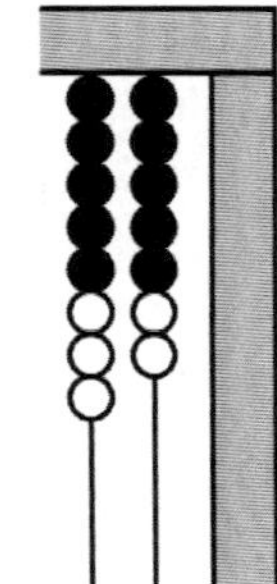

$$\begin{array}{r} 10 \\ -2 \\ \hline 8 \end{array} \qquad \begin{array}{r} \boxed{} \\ -2 \\ \hline 8 \\ \hline 10 \end{array} \qquad \begin{array}{r} 10 \\ -2 \\ \hline 8 \\ \hline 10 \end{array}$$

$$\begin{array}{r} 8 \\ -4 \\ \hline 3 \end{array} \qquad \begin{array}{r} \boxed{} \\ -4 \\ \hline 3 \\ \hline 7 \end{array} \qquad \begin{array}{r} 8 \\ -4 \\ \hline 3 \\ \hline 7 \end{array} \qquad \begin{array}{r} 8 \\ -4 \\ \hline \cancel{3}\;4 \\ \hline 7\;8 \end{array}$$

Find the mistakes by adding.

7	6	1 2	1 3
− 2	− 3	− 6	− 9
4	3	8	6

4 5	5 7	3 9	4 6
2 3	1 7	2 2	4
2 3	4 0	7	3 2

2 5	5 3	6 4	7 5
1 8	2 5	4 9	5 5
1 7	2 8	1 5	3 0

SPECIAL EFFECTS

This section is concerned with some interesting aspects of subtraction, including the relationship between addition and subtraction.

Minuend decreased by 1

Write on the board the following:

$$9 - 4 =$$
$$8 - 4 =$$

Tell the children that the number we start with in subtraction is call the minuend. **What is the 9 called?** [minuend] **What is the other minuend?** [8]

Next ask the children to perform the subtractions on the top two wires. Repeat for 10 – 6, 9 – 6 and for 6 -1, 5 – 1. Ask, **What is special about the pairs of problems?** [The minuend is 1 less in the second problem.] **What do you notice about the remainders?** [The second remainder is 1 less than the first.] Ask them to think of other combinations to try. Then ask them to try to explain why that is true.

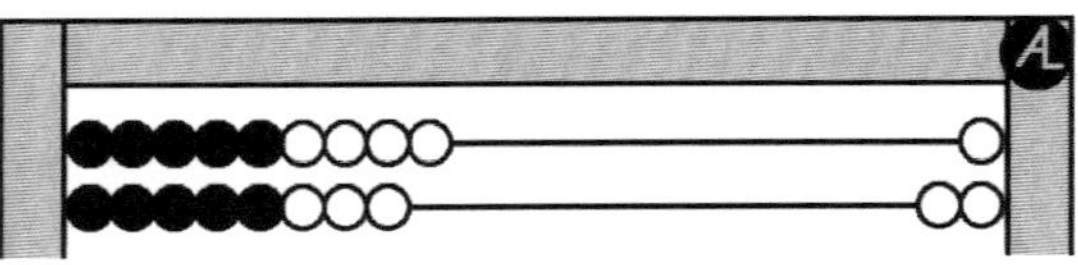

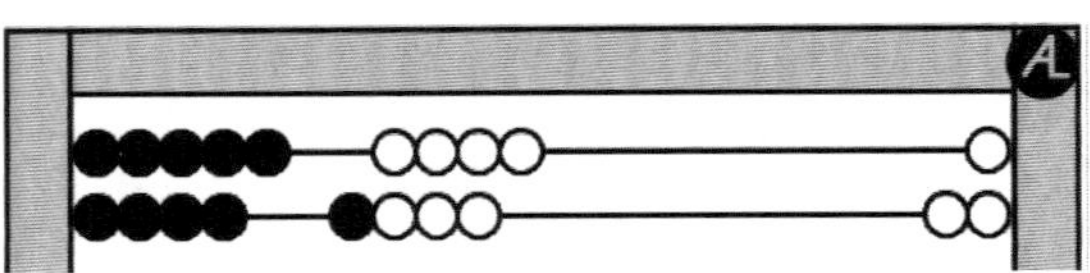

Give them worksheets (5-27 and 5-28) to practice their observations.

THOUGHT QUESTION: What happens to the answer if the minuend is 2 less?

ORAL PROBLEMS. A. Alvaro rode his bicycle 13 blocks. He stopped for a short rest after 8 blocks. How far did he ride after his rest? [5 blocks]

B. The next day Alvaro decided to ride 12 blocks. He stopped at the same place for a rest. How far did he ride after his rest? [4 blocks]

C. Andrea did 11 sit-ups, which was 2 more than her friend did. How many did her friend do? [9 sit-ups]

9 – 4 =			
8 – 4 =			
4 – 1 =			
3 – 1 =			
8 – 2 =			
7 – 2 =			
10 – 3 =			
9 – 3 =			

1 0	9	1 3	1 2
– 6	– 6	– 9	– 9
1 5	1 4	8	7
– 7	– 7	– 6	– 6
1 7	1 6	1 2	1 1
– 8	– 8	– 5	– 5

Subtrahend decreased

The object of this activity is to help the children realize that the answer increases as the subtrahend decreases.

Tell the children that today we will consider what happens when the second number, the subtrahend, is decreased, or gets smaller.

$$9 - 7 =$$
$$9 - 6 =$$

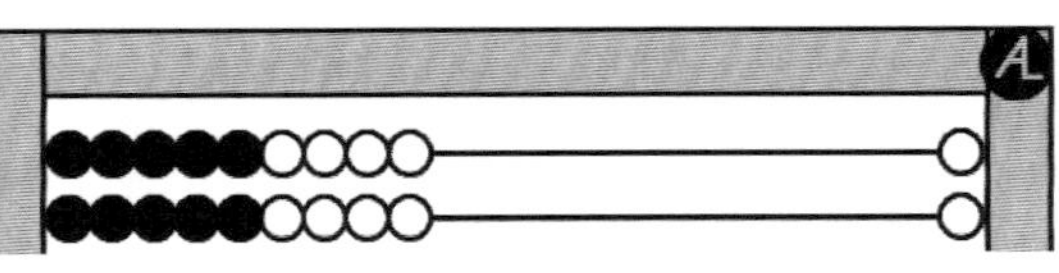

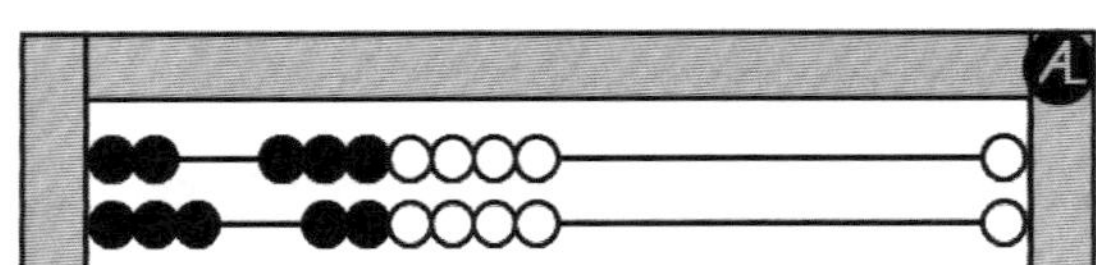

Ask them to enter 9s on the top two wires and to subtract 7 from the first wire and 6 from the second wire. Ask a child to write the answers. Repeat with the pair, 7 – 2 and 7 – 4. Also repeat with 14 – 5 and 14 – 7, but here two wires will be needed for each problem. Ask what happens when the subtrahend is decreased by 1. [The answer is increased by 1.]

The two worksheets (5-29 and 5-30) will also help them see these relationships. When the children have completed them, ask them what happens when the second number, the subtrahend, is 1 less.

ORAL PROBLEMS. A. Max has space for only 13 cars. If he has 19 cars, how many should he give away? [6 cars]

B. Marion's job is to carry out the garbage during the month of June, which has 30 days. If the date is June 23, how many more days does he have left? [7 days]

Counting up

1. Counting up is a good strategy to use whenever two numbers are close in magnitude. Cashiers giving change use; they start with the amount of sale and count up to the amount tended. Most persons also use it to determine how far away March is from July.

Write on the board

$$8 - 6 =$$

and ask the children to enter 6. Then ask them to enter more beads until they have 8 and say, **How many more beads did you enter?** [2] Ask, **Is 8 minus 6 equal to 2?** [yes]

Repeat for $12 - 9$ and $31 - 28$, making sure the lower number is entered first.

A sample worksheet (5-31) is shown.

2. A variation of the counting up strategy is excellent to use for the teen facts. Counting up is used to find how far the lower number is from 10 and that result is added to the ones of the teen number.

Write on the board

$$\begin{array}{r} 14 \\ -8 \\ \hline \end{array}$$

and tell the children enter 8 on their abacuses in the vertical format. Ask, **How many beads do we need to enter to make 10?** [2] **Then how many more do we need to make 14** [4] **So altogether we needed 2 + 4, or 6.** Repeat for $12 - 7$ and $13 - 6$.

ORAL PROBLEMS. A. Kendra wants to be 50 inches tall. If she is 46 inches, how many more inches must she grow? [4 inches]

B. Kurt scored 93 on a 100-point test. How many points did he miss? [7 points]

9 − 7 =	1 2 − 8	1 2 − 7	7 − 5	7 − 4
9 − 6 =				
6 − 3 =	1 3 − 9	1 3 − 8	1 5 − 7	1 5 − 6
6 − 2 =				
12 − 4 =				
12 − 3 =	9 − 6	9 − 5	1 1 − 5	1 1 − 4
10 − 9 =				
10 − 8 =				

8 − 6 =
11 − 8 =
17 − 14 =
20 − 15 =
13 − 10 =
60 − 57 =
42 − 38 =
64 − 59 =
40 − 36 =

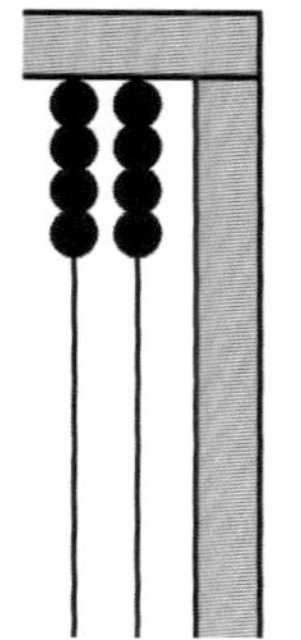

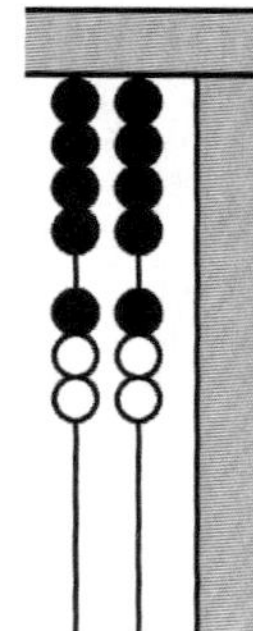

Give them worksheets (5-32 and 5-33) with teens facts for practice. Some might find it helpful to write the number needed to make 10 near the subtrahend, but outside the box. For example, in 14 – 8, a small 2 may be written near the 8 and added to the 4 to find the answer of 6.

ORAL PRACTICE. After the individual work, write the teen facts on the board. If necessary, remind them to quickly add the ones of the minuend with what is needed to make 10 of the subtrahend.

1 4	1 2	1 1	1 7
– 8	– 9	– 6	– 8

1 8	1 1	1 6	1 3
– 9	– 5	– 7	– 6

1 1	1 3	1 2	1 5
– 9	– 8	– 8	– 9

The complement method (optional)

This strategy works in all cases where borrowing is required. Take for example 12 – 8; first subtract backwards, 8 – 2 = 6. Then think of the number needed to complete 10, which is 4.

This complement method of subtraction is also helpful for subtracting 1:00 AM minus 8:00 PM. First subtract backwards, 8 – 1 = 7. This time think of the number needed to complete 12, which is 5.

Write on the board

$$\begin{array}{r} 14 \\ -\ 6 \\ \hline \end{array}$$

and explain that you have a trick that they might want to use. Demonstrate the procedure. First enter the 6. Next subtract the 4 by sliding it towards the top. Finally add what is needed to make 14, which will be what is needed to complete 10, namely 8.

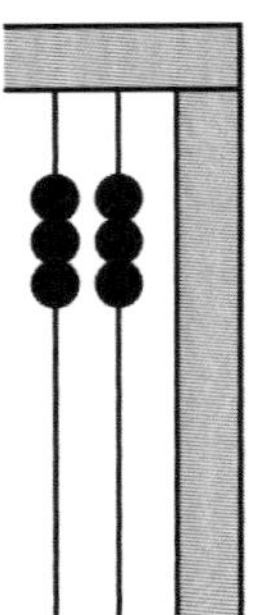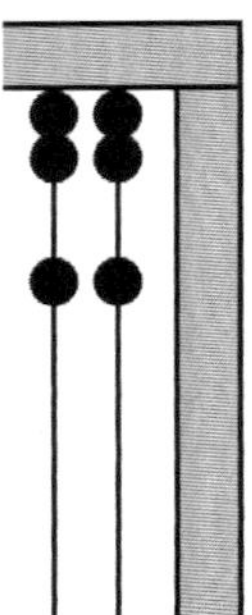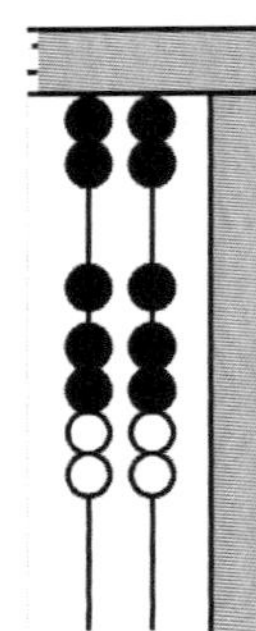

Repeat with 14 – 7 and 12 – 9. Any worksheet with teens will provide practice. After the worksheet, practice can be conducted without the abacus. Ask them to subtract backwards and then to think of the number that is needed to make 10.

ORAL PROBLEMS. A. Marith picked 8 cherry tomatoes one day and 9 the next. How many did she pick? [17 tomatoes]

B. Greg ate 5 of those tomatoes? Now how many were left? [12 tomatoes]

C. Next Lee ate half of the remaining tomatoes. How many did he eat? [6 tomatoes]

D. Carina ate the other half. How many did she eat? [6 tomatoes]

E. How many tomatoes were eaten? [17 tomatoes]

Unit 6
Mastering the subtraction facts

Before the children start to work on mastering the subtraction facts, it is absolutely essential that they thoroughly know the addition facts. This is important for at least two reasons: some of the strategies use adding and, secondly, performing an activity "backwards" is easier if doing it forwards is completely learned.

Here is an example to illustrate the point. A student who wanted to learn how to use the Japanese abacus bought a text. The text taught addition and subtraction together; every new topic was explained for both. After a few weeks of practice, the student became hopelessly confused and set the book and the abacus aside. However, several years later, she picked them up again and this time decided to learn only addition first. No problem. Then she learned subtraction and again, no problem.

Actually it is not necessary to unduly pressure the children to learn these facts in order to proceed to subtraction with several digits. Provide a subtraction table or the Cotter Sum Line, described in Unit 7.

Some of the strategies for learning the subtraction were introduced in the previous unit. But here the facts are studied in groups to facilitate memorization. The strategies include subtracting 1 or 2, subtracting consecutive numbers, subtracting from 11 or 9, subtracting 9 or 8, and the doubles and near doubles. The facts in the teens can also be learned from either counting up or the borrowing trick (10s complement).

SUBTRACTING 1 OR 2

Before starting on the special cases of subtracting 1 and 2, it is good to expand evens + evens = evens and so forth to subtraction.

Subtracting evens from evens...

Review by asking the children, **What do you get when you add an even number to another even number?** [even number] **What about two odd numbers?** [even] **What about one of each?** [odd]

Write on the board

even + even = even	even − even =
odd + odd = even	odd − odd =
even + odd = odd	even − odd =
	odd − even =

For the first case, even − even =, ask them to think of several examples of subtracting two even numbers. [8 − 6, 12 − 4, 10 − 6] Write them on the board in the vertical format. Have the children find the remainders and write them. Then ask, **What happens when you subtract an even number from an even number?** [The remainder is an even number.]

Repeat for the other three cases. You might ask the children to guess the answer first. If the children tire of the activity before it is completed, resume the next day.

even + even = even   even - even = even

odd + odd = even   odd - odd = even

even + odd = odd   even - odd = odd

                                             odd - even = odd

With the answers written, ask the children if they can think of an easy way to remember the results. [If they are the same, the answer is even, if mixed, it is odd.] Finding an easy way to remember is an important skill that is often neglected in teaching.

For practice, give them a page (6-1) of vertical subtraction, but ask them to write an e for even or o for odd beside each number before they perform the subtraction. Then they can do the subtractions to check their work.

ORAL PROBLEMS. A. Tell the children that you are thinking of even numbers whose sum is 8 and whose difference is 4. What are the numbers? [2 and 6]

B. Find two odd numbers whose sum is 12 and whose difference is 6. What are the numbers? [3 and 9]

Subtracting 1 from a number

Subtracting 1 from a number was introduced earlier, but only for numbers 10 or less. Now for practice, it will be expanded to numbers to 100. The vertical format will be used, but the children are to use the wires horizontally.

Write on the board

$$\begin{array}{r} 37 \\ \underline{-\,1} \end{array}$$

Most children will not need their abacuses, but if they do, they can use them with the wires horizontal.

Give them two more practice problems, such as 30 – 1 and 76 – 1 before giving them a worksheet (6-2). As a variation, have them work with partners, reading the answers aloud to each other.

Subtracting 2 from an even number

Review with the children that when 2 is added to an even number, the sum is the next even number. The object of this section is to help the children discover that subtracting 2 from an even number is the preceding even number, and to help them to apply it to appropriate problems.

Ask them to count by 2s backwards from 30. Write several numbers on the board

16   6   38   62   70

and ask them to name the even number before them.

Then write on the board

$$\begin{array}{r} 28 \\ \underline{-\,2} \end{array}$$

and say, **What do you think is the answer?** [26]

If they are unable to give an immediate answer, ask what even number comes before 28? Repeat for 56 – 2 and 40 – 2. The corresponding worksheet (6-3) is shown.

7 0 – 3 0	8 – 4	1 5 – 6	8 – 5
1 1 – 2	6 – 0	7 – 2	1 4 – 9
1 7 – 9	1 2 – 5	1 3 – 7	1 0 – 2

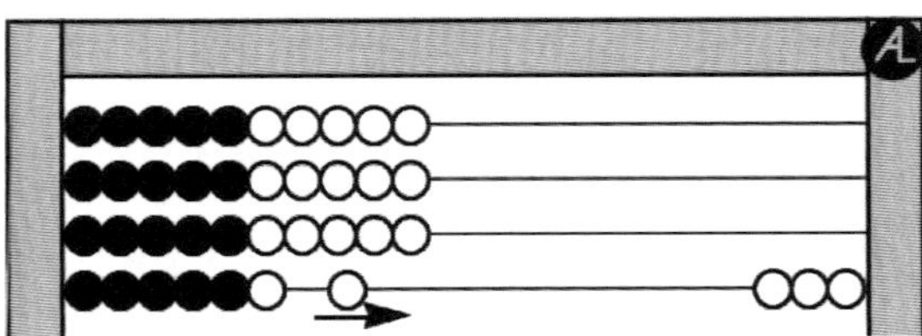

3 7 – 1	4 3 – 1	7 2 – 1	5 6 – 1
8 0 – 1	6 5 – 1	8 – 1	100 – 1
1 4 – 1	5 1 – 1	2 9 – 1	9 0 – 1

2 8 – 2	6 0 – 2	9 2 – 2	4 6 + 2
3 4 – 2	7 8 + 2	5 0 – 2	1 6 – 2
8 2 – 2	100 – 2	1 8 – 2	9 4 – 2

Subtracting 2 from an odd number

Subtracting 2 from a number, either even or odd, could also be thought of as counting back 2.

Review with the children counting by odd numbers both forward and backward. Then write on the board

$$7 \qquad 13 \qquad 59 \qquad 31 \qquad 95$$

and ask the children to name the odd numbers preceding those written.

Next write on the board and say

$$\begin{array}{r} 45 \\ -2 \\ \hline \end{array}$$

Ask the children for the remainder. Repeat for $73 - 2$ and $41 - 2$.

First give them a worksheet (6-4) with only odd numbers minus 2. Later, give them a worksheet (6-5) with even and odd numbers minus 1 or 2.

SUBTRACTING CONSECUTIVES

These activities are special cases of the adding up and the borrowing trick.

Teach the children that **consecutive** numbers are numbers that follow each other in order. The numbers 5 and 6 are consecutive, but 10 and 13 are not. Ask, **Are 7 and 10 consecutive?** [no] **Are 17 and 18?** [yes]

Subtracting consecutive ones

Subtracting consecutive ones where the minuend is greater than the subtrahend can most easily be solved by adding up. Write on the board

$$8 - 7 = \qquad 3 - 2 = \qquad 9 - 8 =$$

Call on various children to write the remainders. [1]

Next write on the board

$$17 - 8 = \qquad 12 - 3 = \qquad 18 - 9 =$$

and ask the children how these are different from the previous row. [The subtrahend is greater than the ones in the minuend.] Ask, **What is the answer to these problems?** [9] Cover up the 1 on any of the problems and ask them if they can do the subtraction. [no]

Finally, give them a mixture of the problems

$$13 - 2 = \qquad 15 - 6 = \qquad 25 - 6 =$$

before giving them worksheets (6-6 and 6-7) as shown.

Subtracting consecutive evens or odds

This work is similar to that with the consecutive ones, but now the emphasis is on consecutive evens or odds. Write on the board

$$6 - 4 = \qquad 8 - 6 = \qquad 7 - 5 =$$

Ask the children the answers. [2] Write on the board

$$14 - 6 = \qquad 16 - 8 = \qquad 15 - 7 =$$

and ask the children, **If the answers to the first row is 2, what must be the answers to the second row?** [8]

Give them worksheets (6-8 to 6-10) similar to those shown.

12	−	4	=
14	−	2	=
13	−	1	=
16	−	8	=
15	−	7	=
18	−	6	=
17	−	9	=
19	−	7	=
16	−	4	=

| 57 | 88 | 23 | 30 |
| − 9 | − 6 | − 5 | − 2 |

| 43 | 74 | 17 | 92 |
| − 1 | − 6 | − 5 | − 4 |

| 36 | 25 | 59 | 61 |
| − 8 | − 7 | − 7 | − 3 |

| 65 | 38 | 51 | 97 |
| − 7 | − 6 | − 3 | − 6 |

| 87 | 71 | 46 | 34 |
| − 9 | − 2 | − 7 | − 6 |

| 14 | 23 | 58 | 66 |
| − 2 | − 5 | − 9 | − 8 |

SUBTRACTING FROM 11 OR 9

Subtracting from 11 or 9 presents no new techniques. It is the combination of the facts that make 10 combined with what happens when the minuend is increased or decreased by 1.

Subtracting from 11

Write on the board

$$10 - 4 =$$
$$11 - 4 =$$

and tell the children that today we will concentrate on the 11 facts. Ask the children what is 10 − 4. Then ask what 11 − 4 must be.

Demonstrate separating 11. Enter 11, separate 2, and ask the children to recite the fact. [11 − 2 = 9] Separate another bead and ask for the fact. [11 − 3 = 8] Continue to 11 − 9. Have the children work with partners separating 11 and saying the facts.

ORAL PRACTICE. Write the 11 facts on the board in random order in either format and ask the children to recite the remainders.

Give them written practice (6-11 and 6-12) with the facts and with the facts in the higher decades as shown.

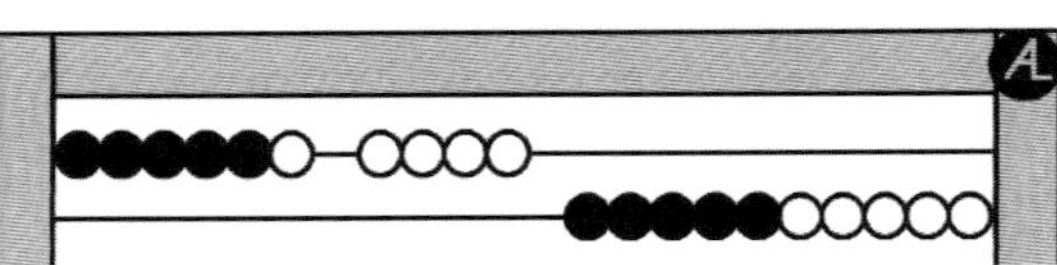

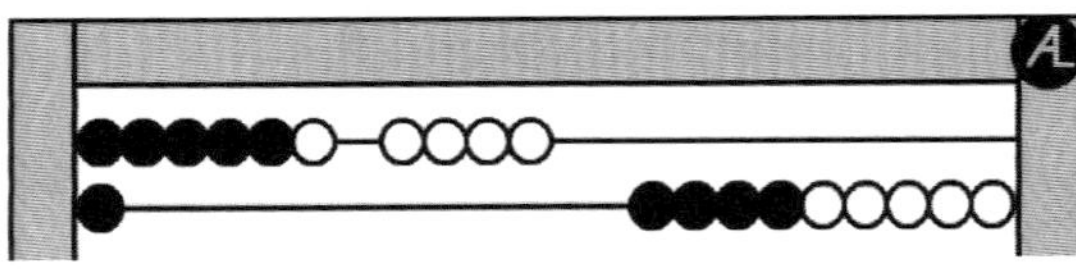

11	−	4	=
11	−	7	=
11	−	9	=
11	−	3	=
11	−	6	=
11	−	2	=
11	−	5	=
11	−	8	=
11	−	6	=

| 31 | 81 | 41 | 91 |
| − 4 | − 3 | − 5 | − 8 |

| 11 | 61 | 71 | 51 |
| − 9 | − 7 | − 1 | − 4 |

| 21 | 41 | 61 | 31 |
| − 2 | − 0 | − 6 | − 5 |

Subtracting from 9

Although learning the 9-facts is harder than the 11-facts, the application in higher decades is easier because no borrowing is involved.

Write on the board

$$10 - 3 =$$
$$9 - 3 =$$

and ask the children to enter 10 and 9 on the top two wires and subtract 3 from both. Then ask, **We know what 10 – 3 is; what happens when 3 is subtracted from 9?** [one less] Repeat for other subtrahends until they are certain of the result.

ORAL PRACTICE. Write the facts on the board and ask the children to recite the remainder as you point to one.

9	9	9	9	9	9	9
−4	−8	−2	−5	−3	−7	−6

For written practice give them the worksheets (6-13 to 6-16) shown, including one that asks for subtraction from 9, 10, and 11.

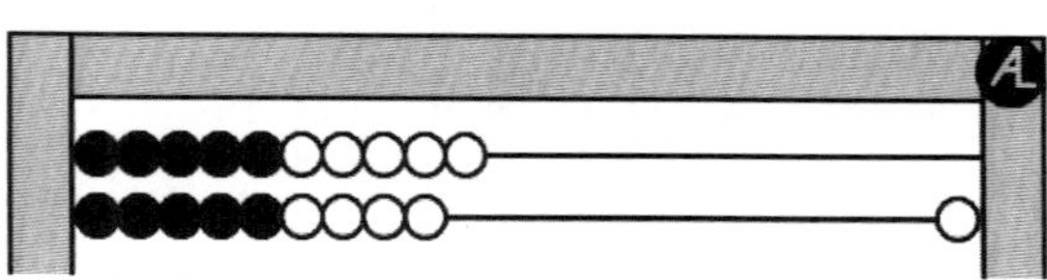

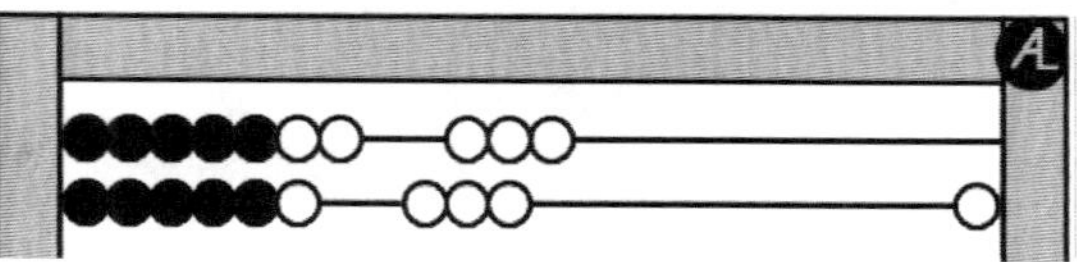

9 − 3 =		5 9	8 9	6 9	3 9
9 − 7 =		− 7	− 5	− 4	− 6
9 − 6 =					
9 − 1 =		7 9	1 9	4 9	2 9
9 − 5 =		− 2	− 8	− 7	− 1
9 − 8 =					
9 − 2 =		9 9	5 9	9	2 9
9 − 4 =		− 3	− 9	− 0	+ 6
9 − 9 =					

SUBTRACTING 9 OR 8

The objective is that the children will be able to demonstrate the strategy of subtracting 9 or 8 by first subtracting 10 and then adding 1 or 2, respectively.

9 − 3 =		9	1 1	1 1	9
11 − 8 =		− 4	− 5	− 7	− 6
10 − 1 =					
11 − 5 =		1 1	9	1 1	6
9 − 7 =		− 3	− 8	− 6	− 5
10 − 4 =					
9 − 6 =		9	9	1 1	1 1
11 − 9 =		− 2	− 3	− 8	− 4
11 − 2 =					

Subtracting 9 from a number

Write on the board

$$17 - 10 = \qquad 14 - 10 = \qquad 12 - 10 =$$
$$17 - 9 = \qquad 14 - 9 = \qquad 12 - 9 =$$

and tell the children that today they will work on subtracting 9 from a number, which is much like subtracting 10.

Ask the children to enter 17 on their abacuses. Tell them to subtract the 10 from the top wire; ask one of them to write the answer. [7] Ask them enter the 17 again. To subtract the 9, ask them to remove the 10 and give back a 1. **Does it look like we subtracted 9?** [yes] **Now what is the remainder?** [8] Repeat for the remaining examples.

Next give them 35 – 9. Ask them to use the rule of subtracting 10 and adding a 1. Encourage them to use their abacuses to see if they are right. The children could work with partners, one subtracting with the new rule and the partner, using another method.

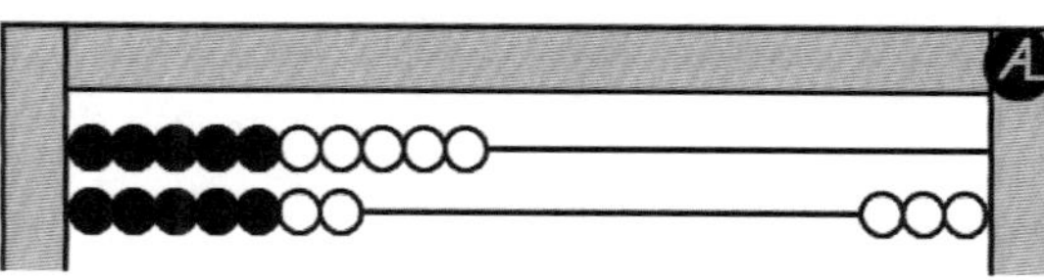

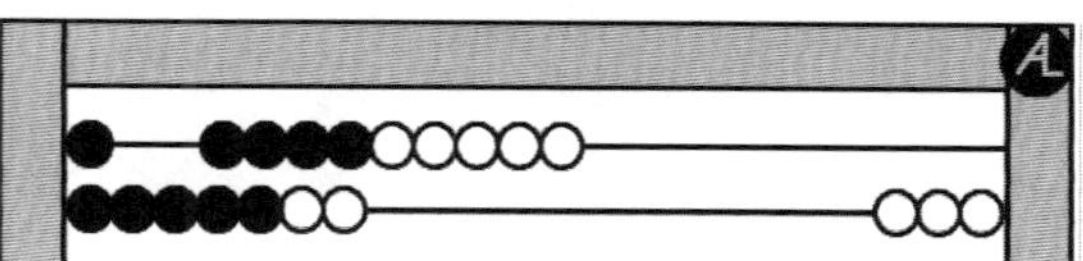

ORAL PRACTICE. Write the facts for the children to practice with partners. Include some problems like 76 – 9, 28 – 9, and 36 – 9.

Give them worksheets (6-17 and 6-18), one with the facts and one with higher numbers.

Subtracting 8 from a number

Write on the board

$$15 - 10 = \qquad 17 - 10 = \qquad 13 - 10 =$$
$$15 - 8 = \qquad 17 - 8 = \qquad 13 - 8 =$$

Remind the children that they already know how to subtract 9 from a number, and ask, **How do you think we can subtract 8 from a number?** [Subtract 10 and add 2.]

Ask the children to enter 15. To subtract the 8, ask them to remove the 10 and give back 2. **Does it look like we subtracted 8?** [yes] **Now what is the remainder?** [7] Repeat for the remaining examples.

ORAL PRACTICE. Write the facts for the children to practice. Include some problems like 54 – 8, 37 – 8, and 42 – 8.

Give the children a variety of worksheets (6-19 and 6-22) as shown.

17	– 9	=
13	– 9	=
18	– 9	=
12	– 9	=
15	– 9	=
16	– 9	=
14	– 9	=
11	– 9	=
10	– 9	=

44	– 9	=
86	– 9	=
67	– 9	=
81	– 9	=
32	– 9	=
16	– 9	=
51	– 9	=
75	– 9	=
33	– 9	=

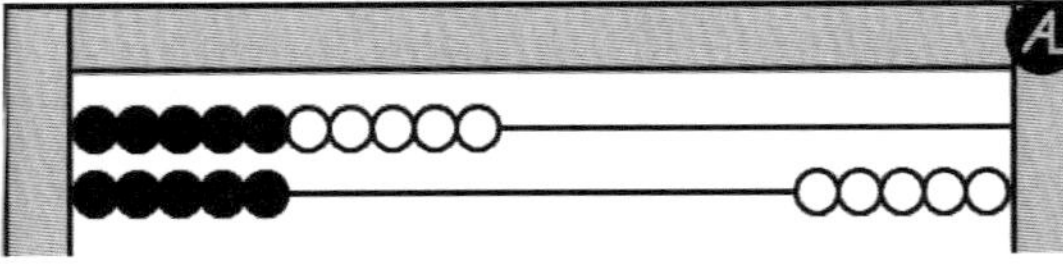

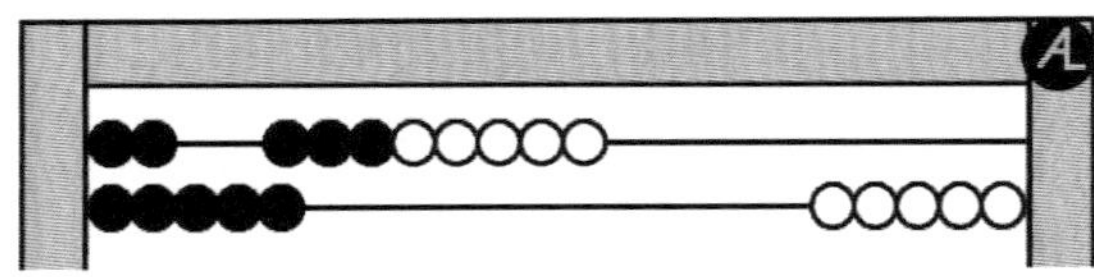

15	– 8	=
13	– 8	=
17	– 8	=
14	– 8	=
12	– 8	=
16	– 8	=
11	– 8	=
10	– 8	=
17	– 8	=

56	– 8	=
38	– 8	=
77	– 8	=
62	– 8	=
83	– 8	=
24	– 8	=
95	– 8	=
49	– 8	=
31	– 8	=

12	– 9	=
11	– 8	=
15	– 8	=
10	– 9	=
18	– 9	=
13	– 8	=
17	– 9	=
16	– 9	=
14	– 8	=

1 3	1 4	1 5	1 7
– 8	– 9	– 9	– 8
1 2	1 8	1 6	1 3
– 8	– 9	– 8	– 9
1 4	1 7	1 6	1 5
– 8	– 9	– 9	– 8

THE DOUBLES GROUPS

Of the doubles groups, only the doubles and near doubles, not the middle doubles, will be studied in subtraction. These two groups include four facts below 10 and four facts above 10 that are not covered by any other strategies. They are 8 – 4, 7 – 3, 7 – 4, 6 – 3, 14 – 7, 13 – 6, 13 – 7, and 12 – 6.

Doubles

The doubles groups are also important to learn from the standpoint of fractions. The concept of one-half will be introduced.

Place 6 identical objects on a table where all the children can see. Call up two children and ask them to divide the objects in HALF so that each receives the same amount. Say, **So one half of 6 is what?** [3] Repeat with 8 objects and two different children.

With the wires vertical, ask the children to enter 2. Ask them to cover either half and recite the fact. [2 – 1 = 1] Next they are to enter 2 more, cover either half and recite that fact. [4 – 2 = 2] Continue to 18. If desired, the children can write these facts.

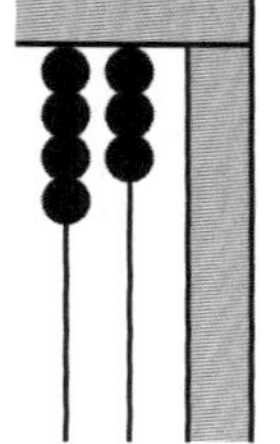

ORAL PRACTICE. Write the even numbers in random order on the board, 8, 6, 12, 14, 8, 10, 2, 18, 16, 4, and ask the children to name half of each number.

Or, give each group of two or three children a set of numbers 1 to 9 and a set of evens 2 to 18. Ask them to lay out the evens and match its half with the other cards. Finally ask, **What is the answer when you subtract half of a number?** [The same number, or the other half.]

Then give them a worksheet (6-23).

Write one half of each number.			
12_	6_	16_	2
8_	10_	2_	– 1
14_	4_	18_	
1 2 – 6	1 6 – 8	1 4 – 7	4 – 2
6 – 3	1 8 – 9	1 0 – 5	8 – 4

Near doubles

Just as every even number has a double, every odd number has a near double. Ask the children to enter 7 on their abacuses with the wires vertical. Call on them to cover each side and recite the corresponding facts. [7 – 3 = 4 and 7 – 4 = 3] Repeat for 13 [13 – 6 = 7 and 13 – 7 = 6] and 9. [9 – 4 = 5 and 9 – 5 = 4]

ORAL PRACTICE. Write the odd numbers from 3 to 17 and ask the children to practice saying the associated facts: 5   11   3   9   17   13   9   7   15.

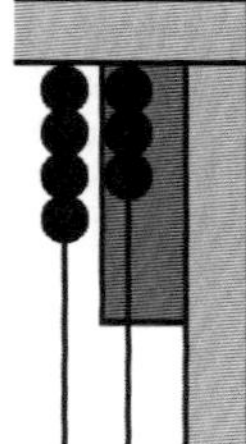
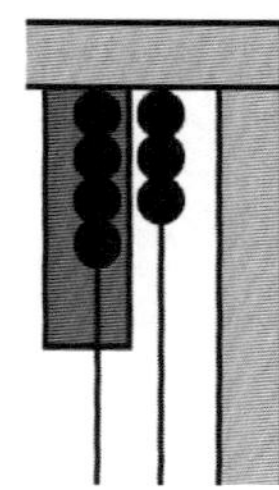

For practice give them the worksheets (6-24 to 6-26) shown. The second worksheet gives practice with both the doubles and near doubles while the third worksheet gives practice in the higher decades.

7	1 3	1 1	1 5
- 3	- 6	- 5	- 7

9	1 5	1 1	1 7
- 5	- 8	- 5	- 8

5	9	7	1 7
- 3	- 4	- 4	- 9

13 – 6 =		19 – 4 =
8 – 4 =		87 – 3 =
11 – 5 =		68 – 4 =
5 – 2 =		53 – 7 =
12 – 6 =		32 – 6 =
7 – 4 =		21 – 5 =
15 – 8 =		43 – 6 =
6 – 3 =		34 – 7 =
13 – 7 =		89 – 5 =

Unit 7
Adding and subtracting in the thousands

Although mastery of the addition and subtraction facts is highly desirable, it is not an absolute prerequisite for adding and subtracting multi-digit numbers. Additional work through games (see *Math Card Games*) can take place concurrently with mastery of the algorithms. This work with larger provides motivation to learn the facts and practice in using strategies.

To add these larger numbers, the abacus will be used with a different format, more like the traditional Japanese or Chinese abacuses except that two wires, instead of one, will represent each number. This work will enable the children to discover for themselves the algorithms for addition and subtraction.

The Base 10 Picture Cards, with a picture of a single cube to represent 1, a row of 10 cubes to represent 10, a hundred square to represent 100, and a thousand cube to represent 1000, provide an important transition from side 1 to side 2 of the abacus. They help children understand that 10 ones = 1 ten, 10 tens = 1 hundred, and 10 hundreds = 1 thousand.

The subtraction algorithm used in this unit is a newer, simpler algorithm, which does not call for crossing out any numbers and starts at the left. It allows any digit in the remainder to be calculated without going through the entire procedure. However, the traditional algorithm works on the abacus as well.

TABLES

In today's world, information is often arranged in tables. The addition and subtraction tables provide an excellent introduction; a copy of each is found in the appendix. The addition table was introduced in Chapter 4. Tables also help the children organize their memorization of facts.

Addition table

The addition table can be used in several ways. Ask the children to figure out how to use it to find the sums, if they do not already know. For example, to find 5 + 6, first find the column with 5 along the top using the right index finger. Next find the row with 6 along the left side using the left index finger. Move the right finger down in the column until it is in the same row as the 6. Read the sum as 11.

Give the children several facts to find with the table: 7 + 6 and 9 + 8. Let them work with partners thinking of facts and finding the sums.

+	1	2	3	4	5	6	7	8	9
1	2	3	4	5	6	7	8	9	10
2	3	4	5	6	7	8	9	10	11
3	4	5	6	7	8	9	10	11	12
4	5	6	7	8	9	10	11	12	13
5	6	7	8	9	10	11	12	13	14
6	7	8	9	10	11	12	13	14	15
7	8	9	10	11	12	13	14	15	16
8	9	10	11	12	13	14	15	16	17
9	10	11	12	13	14	15	16	17	18

To further help the children gain an appreciation of the table, ask these and similar questions:

A. According to the table, how many combinations have 4 as a sum? [3] Name them. [1 + 3, 2 + 2, 3 + 1]

B. What sum has the most combinations? [10]

C. What sums have the fewest combinations? [2, 18]

D. Name all the ways to make 9. [1 + 8, 2 + 7, 3 + 6, 4 + 5, 5 + 4, 6 + 3, 7 + 2, 8 + 1]

E. Where are the doubles located on the table? [On the diagonal from the upper left to the lower right]

F. Color all the squares with even numbers. What do you get? [a checkerboard]

G. Find some addition twins and color each set a different color. What do you get? [a mirror design about the evens diagonal, that is, a symmetric design about the line from the upper left corner to the lower right corner]

H. Invite the children to make observations.

Subtraction table

Below are two types of subtraction tables. In the table on the left the facts requiring borrowing are highlighted by showing both the minuends and remainders in italics. The squares that could be either 0 or 10 are blank.

	1	2	3	4	5	6	7	8	9	10	11	12	13	14	15	16	17	18
−1	0	1	2	3	4	5	6	7	8	9								
−2		0	1	2	3	4	5	6	7	8	9							
−3			0	1	2	3	4	5	6	7	8	9						
−4				0	1	2	3	4	5	6	7	8	9					
−5					0	1	2	3	4	5	6	7	8	9				
−6						0	1	2	3	4	5	6	7	8	9			
−7							0	1	2	3	4	5	6	7	8	9		
−8								0	1	2	3	4	5	6	7	8	9	
−9									0	1	2	3	4	5	6	7	8	9

− or	1 11	2 12	3 13	4 14	5 15	6 16	7 17	8 18	9	10
1		1	2	3	4	5	6	7	8	9
2	*9*		1	2	3	4	5	6	7	8
3	*8*	*9*		1	2	3	4	5	6	7
4	*7*	*8*	*9*		1	2	3	4	5	6
5	*6*	*7*	*8*	*9*		1	2	3	4	5
6	*5*	*6*	*7*	*8*	*9*		1	2	3	4
7	*4*	*5*	*6*	*7*	*8*	*9*		1	2	3
8	*3*	*4*	*5*	*6*	*7*	*8*	*9*		1	2
9	*2*	*3*	*4*	*5*	*6*	*7*	*8*	*9*		1

To use the tables, first search for the minuend along the top row(s) above the heavy line. Next find the row with the subtrahend along the leftmost column. The remainder is found at their intersection.

Help the children see interesting aspects of the tables, as well as review strategies, by asking questions such as:

A. Are there more facts with borrowing or without? [without borrowing]

B. How many numbers can be subtracted from 4? [three: 1, 2, 3]

C. How many numbers can be subtracted from 8? [seven: 1, 2, 3, 4, 5, 6, 7, (ignoring 0s)]

D. Find and write all the facts with 5 as a subtrahend. [5 − 5, 6 − 5, 7 − 5, 8 − 5, 9 − 5, 10 − 5, 11 − 5, 12 − 5, 13 − 5, 14 − 5] What happens to the remainders? [They increase as the minuend increases.]

E. Find and write all the facts with 9 as a remainder. [11 − 2, 12 − 3, 13 − 4, 14 − 5, 15 − 6, 16 − 7, 17 − 8, 18 − 9] Do you see a pattern? [The subtrahend is one more than the ones of the minuend.]

G. Find and write all the facts with 5 as a remainder. [6 − 1, 7 − 2, 8 − 3, 9 − 4, 10 − 5] Do you see a pattern? [Both the minuend and subtrahend increase by 1.]

H. Which chart do you like better?

THOUGHT QUESTION: Why is 19 not on the table? [All the numbers from 1 to 9 can be subtracted from 9, so no borrowing is necessary.]

COTTER SUM LINE

The Cotter Sum Line is a special number line that gives all the addition and subtraction facts simply and quickly.

$$\begin{array}{ccccccccccc} 0 & 1 & 2 & 3 & 4 & 5 & 6 & 7 & 8 & 9 \\ \hline 0 & {}^{1}\ 2 & {}^{3}\ 4 & {}^{5}\ 6 & {}^{7}\ 8 & {}^{9}\ 10 & {}^{11}\ 12 & {}^{13}\ 14 & {}^{15}\ 16 & {}^{17}\ 18 \end{array}$$

The sum of two numbers is found by locating the addends above the line, finding their middle, and reading the sum directly below. For example, find 8 + 8: Since the middle is 8, the sum of 16 is read directly below. All doubles are read directly below.

Next try 2 + 8. First find the middle by placing your index fingers on the 2 and the 8 above the line and letting your fingers hop towards each other until they meet at 5. The sum is then read as 10.

For two numbers without a middle, the sum is read as the number between them. For example, 6 + 7 is 13. To find 2 + 7, start at 2 and 7. After hopping, the fingers will end at 4 and 5, where the sum is read as 9.

The Sum Line can also be used to show all the ways to equal, for example, 9. Start with the 9 below the line. The first two addends are 4 and 5. The first hop outward yields 3 and 6; the next hop, 2 and 7; then 1 and 8; and finally 0 and 9.

Subtraction is the same as finding all the ways to equal a number, but hopping is stopped when a finger lands on the desired quantity being subtracted. For example, find 15 − 6. Start at 15. The initial numbers are 7 and 8. Hop outward to 6 and 9. Since we wanted 15 − 6, the remainder is 9.

HUNDREDS AND THOUSANDS

Children need to experience the quantities of 100, 200, and so forth separately before combining with tens and ones to form numbers like 582. The AL abacus with 100 beads can be used for this as well as Base 10 Picture Cards, shown simplified at the left.

Thousand   Hundred   Ten   One

The hundreds

1. Ask the children, **How many beads are on an abacus?** [one hundred] Hold up two abacuses and ask, **How many beads do you think are on two abacuses?** [two hundred]

Say, **You know how we write one hundred; how do you think we write two hundred?** Show the place value card of 200 found in the appendix. Point to the 2 while saying **two**, to the middle 0 while saying **hun-**, and the last 0 while saying **-dred**. Repeat for quantities 300 and 400.

Place 100 identical items in a container; show it to the children and tell them that it is 100. Then show them containers with 200, 300, or other multiples of 100 and ask the children to guess the quantities.

2. Hold up the various hundred cards one at a time and ask the children to read them. Also ask them to show the quantities. If extra abacuses are unavailable, use copies of the "abacus" cards, found in the appendix, to represent one hundred. Along with the hundred cards, hold up ten cards and one cards for them to read.

3. Show a quantity such as 245 by setting up two abacuses, each showing 100, and a third abacus showing 45. Pick up the 200, the 40, and the 5 cards; combine them by overlapping to make 245. Provide each child or group of children a set of place value cards. Give them other combinations to construct with abacuses and place value cards, such as 381 and 179.

4. Comparison of two quantities can help the children appreciate the relative magnitudes. Write on the board

321  123

Have the children work with a partner. Give each child an abacus and 9 abacus cards; each pair also needs a "greater than or less than" card. The child on the left enters the left quantity while the child on the right enters the quantity on the right. Then they decide which quantity is greater and place the sign accordingly.

Give them two examples

452 ◯ 402   512 ◯ 517

before giving them a worksheet (7-1) to be done with a partner. After correctly placing the < or > sign, they can take turns copying it on to the worksheet.

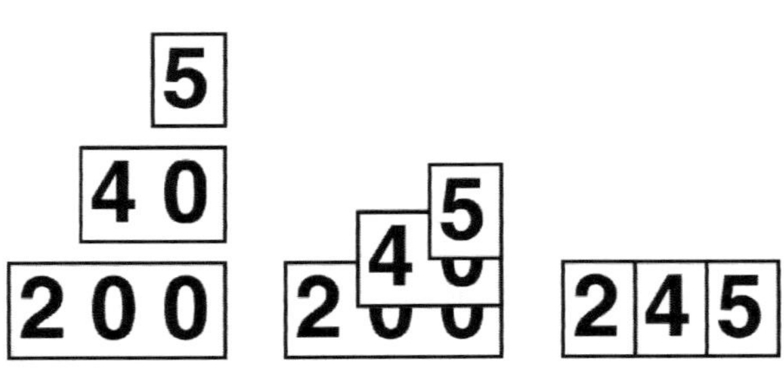

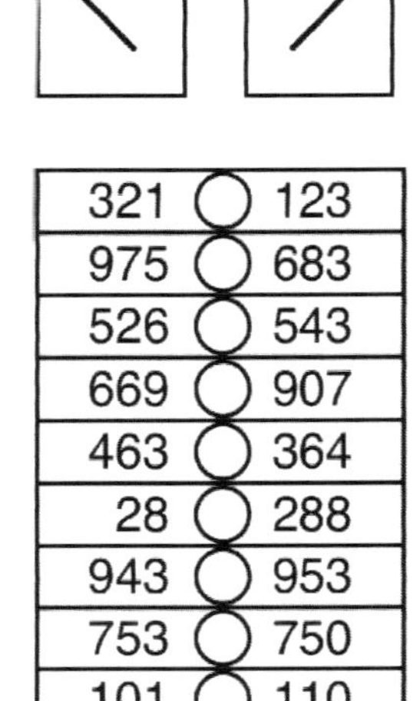

321	◯	123
975	◯	683
526	◯	543
669	◯	907
463	◯	364
28	◯	288
943	◯	953
753	◯	750
101	◯	110

FUTURE REVIEW. After the children have worked with the < and > signs for the hundreds, ask them if they figured out the rules to do the work without the abacuses. Ask, **Which number do you look at first?** [the hundreds] Then ask, **If the hundreds are the same, what do you look at next?** [the tens] Lastly ask, **If both the hundreds and the tens are the same, what do you do?** [look at the ones]

Write a few sets on the board for them to do without abacuses. Discuss following the rules.

Adding with sums in the hundreds

1. The purpose of this activity is to introduce carrying with sums beyond one hundred. It is not necessary at this point that the children start with the ones.

Write on the board

$$\begin{array}{r} 87 \\ + 45 \\ \hline \end{array}$$

and ask the children to try it on an abacus. Then ask, **What is the problem?** [The sum will not fit on one abacus? Use two abacuses with the second placed directly below the first. Start the addition by entering 8 tens on the first abacus and 7 ones near the bottom of the second abacus. Next enter the 4 tens by filling the first abacus and continuing into the second. Finally enter the 5 ones and construct the sum 132 with the place value cards.

Let the children work in pairs to find the sums for 75 + 52, 97 + 68, and 45 + 86. Then give them worksheets (7-2).

ORAL PROBLEMS. A. A ball costs 69ç. Tou wanted two balls; what would they cost? [$1.38]

Now is a good time to explain that a DOLLAR is 100 cents, so the answer would be 1 dollar and 38 cents.

B. Ser had 300 pennies. How many dollars is this? [3]

C. Mee bought a scarf at a sale for 49ç. She also bought some buttons for 75ç. What did the two items cost? [$1.24]

D. Amanda bought two gifts, one costing 86ç and the other 55ç. What was the total cost? [$1.41]

E. Vang had 23 days of vacation from school. Eighteen days have gone by. How many days does he have before school starts again? [5 days]

F. Jonathan had saved $13 for a special gift that cost $20. How much more did he have to save? [$7]

G. David walked for 28 minutes on Monday, 35 minutes on Wednesday, and 37 minutes on Friday. How many minutes did he walk that week? [100 minutes]

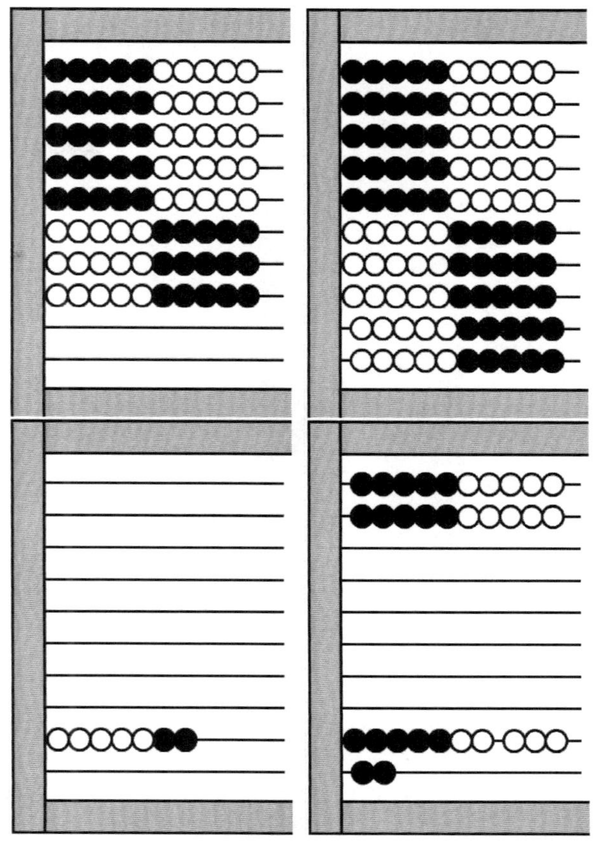

87	63	26
+ 45	+ 89	+ 74
37	58	47
+ 69	+ 94	+ 63
52	73	68
+ 76	+ 28	+ 39

2. Now tell the children that they are able to add numbers in the hundreds. They will each need 9 abacus cards.

Write on the board

$$\begin{array}{r} 376 \\ +\,253 \\ \hline \end{array}$$

and start as before. First enter 376 and then add 253 as shown.

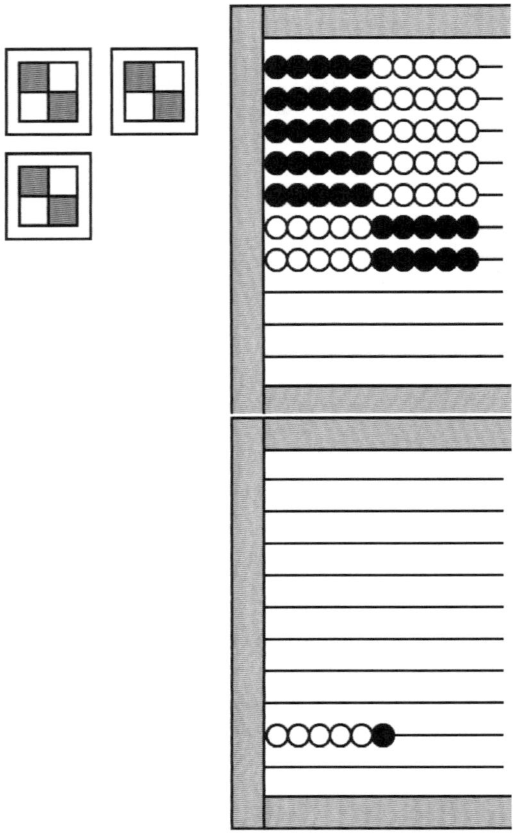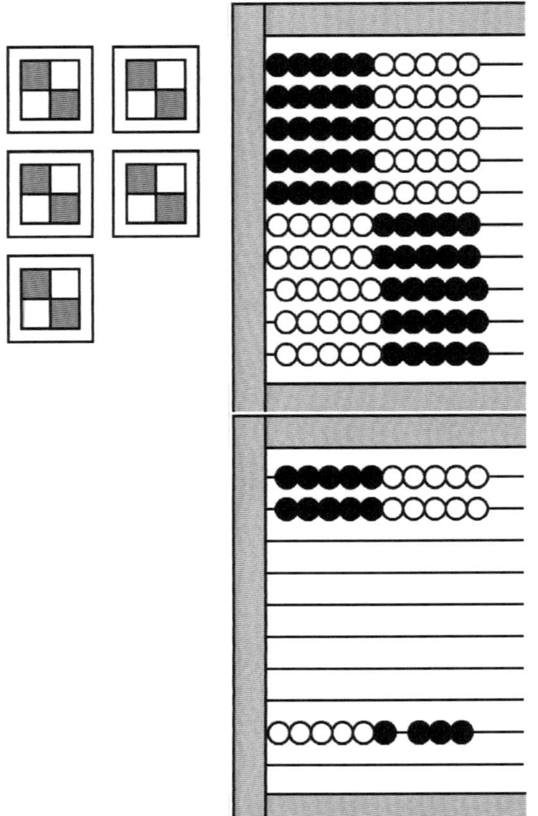

The hundreds cards should be placed in the even-odd format for easy recognition and in preparation for the next step on the abacus.

Help the children construct the sum with the place value cards by asking, **How many hundreds do you have?** [6] Be sure they count the abacus with 100 beads. (Alternately the abacus with the 100 beads could be traded for an abacus card) **How many tens?** [2] **How many ones?** [9] Using the place value cards prevents the children from writing the answer as 5129.

Ask the children to do these examples:

$$\begin{array}{r} 258 \\ +\,139 \\ \hline \end{array} \qquad \begin{array}{r} 682 \\ +\,226 \\ \hline \end{array}$$

376	389	508
+253	+442	+277
754	660	348
+186	+291	+249
567	243	197
+ 33	+598	+455

before giving them a worksheet (7-3) for practice.

ORAL PROBLEM. A year has 365 days. How many days are there in two years? [730 days]

The thousands

1. Tell the children that today they will learn about the thousands. Review by asking, **What do we call 10 ones?** [10] **What do we call 10 tens?** [100]

If available, make a stack of ten abacuses while counting one hundred, two hundred, and so on to ten hundred. Upon reaching 10 hundred, tell the children, **Ten hundred has a special name; it is called one THOUSAND.** Show the children the abacus-cube (cutout is found in the appendix) and tell them that they will use it for one thousand.

Hold up for the children the place value card for 1000 and tell them that that is how we write one thousand. Hold up the cards for 2000, 4000, 7000, 800, 20, 6000 and so forth and ask the children to read them and to show the quantities with abacuses, abacus cards, and the abacus cubes.

2. Form 2867 with two cubes, 8 abacus cards, and 67 on an abacus. Ask the children to construct the quantity with the place value cards and to read the number. Give them other quantities, 4602, 3841, 2045, and 4009, to construct and read. Be sure the word "and" is not said while reading the number.

3. Write on the board the following numbers:

<pre>
4027 _____
7935 _____
 290 _____
4083 _____
</pre>

Tell the children that we want to put these in order from the greatest to the least. Ask a child to name the greatest number. [7935] That number is then crossed out and written on the top line. Ask another child now to find the greatest and to write it on the next line. [4083] Continue until all the numbers are in order.

The children could work in groups with a worksheet (7-4) forming, constructing, and reading quantities.

4. Write on the board:

<pre>
1200
1400
3600
</pre>

and tell the children that people read them two different ways. Help them discover that 1200 is either one thousand two hundred or twelve hundred. Ask them if they can think of any instances where the twelve hundred would be used. [house numbers and the current year] Give them other numbers to read both ways: 1300, 2400, 8200, 1645, and 1989.

If they are ready you might tell them that sometimes we even skip the word hundred, as in 1968 read as nineteen sixty-eight.

THOUGHT QUESTION. How many tens are in 1000? [100]

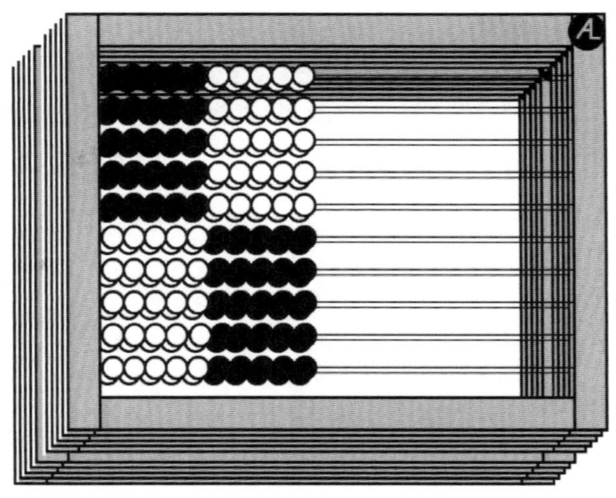

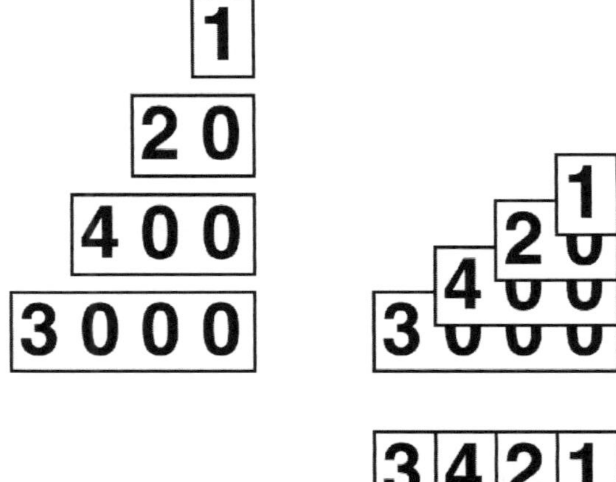

Compose.	Put in order.	
6371	4487	
3450	56	
9003	560	
646	3	
4100	506	
1080	4478	
930	8009	
2064	74	

Adding in the thousands

Tell the children they know how to add numbers in the hundreds. Now they are ready to add numbers in the thousands. Write on the board

$$\begin{array}{r} 1489 \\ + 756 \\ \hline \end{array}$$

Ask a child to construct the number 1489 with the place value cards and another child to form the number with abacuses, cards, and cubes. See the first figure. For a child having difficulty representing the number, separate the place value cards and ask the child to form each component separately.

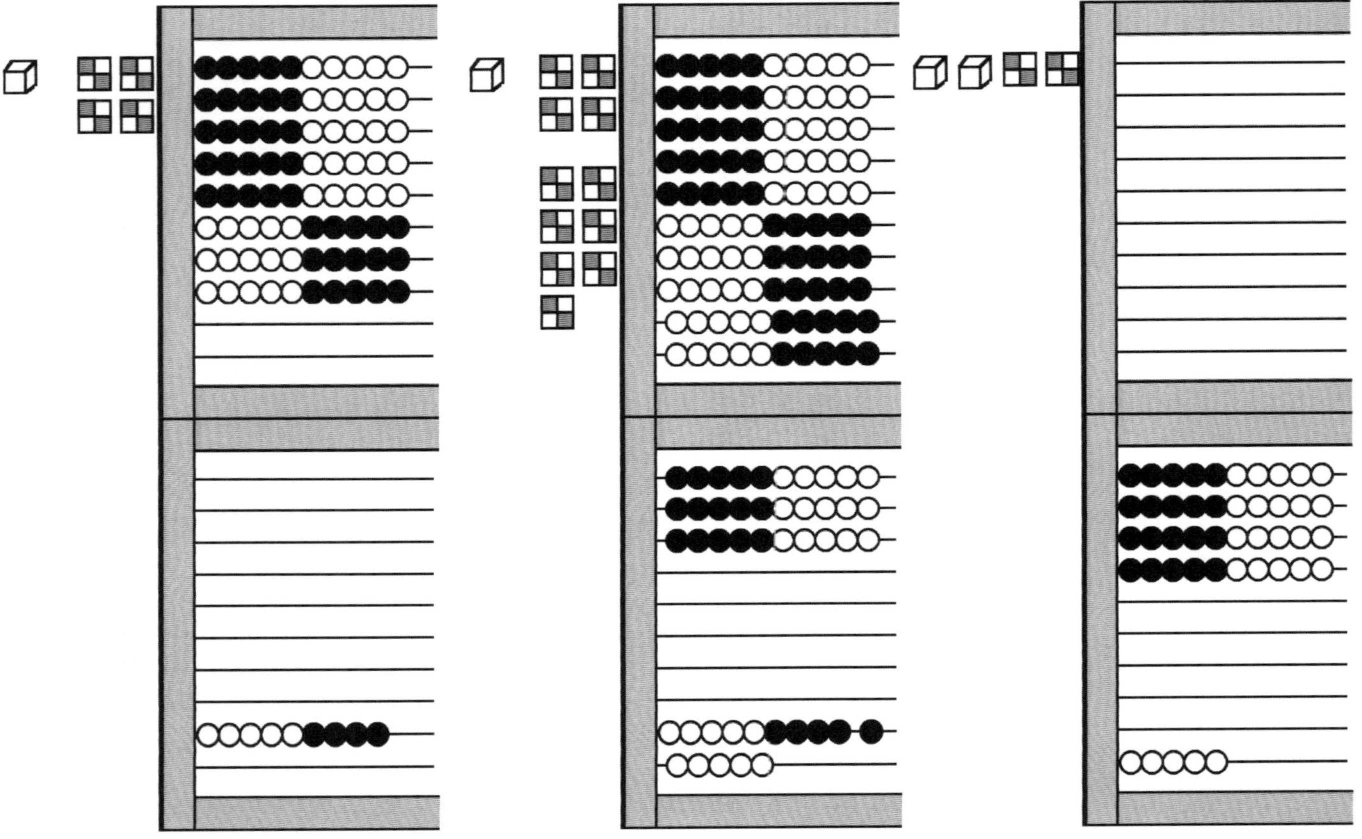

Next add the 6 ones, 5 tens, and 7 hundreds; see the second figure. In the last figure, the new ten is moved to the other tens and the 100 on the abacus is replaced by another abacus card, both optional. Also, the ten abacus cards are traded for an abacus cube. The final answer can be constructed with the place value cards and written.

Break the children into groups and ask them to add 2476 + 1658 and to add 3091 + 989. Then give them worksheets (7-5 and 7-6).

1489	805
+ 756	+ 758
2377	2473
+ 816	+ 1489
3845	476
+ 323	+ 2631

Subtracting in the thousands

Once the children are very familiar with addition using the
concrete representations, they are ready to work with subtraction. Write on the board

$$\begin{array}{r} 4567 \\ -\,2819 \\ \hline \end{array}$$

Ask a child to construct the number 4567 with the place
value cards and another child to form the number with
abacuses, cards, and cubes. See the first figure.

Start by first subtracting 2 thousand. Ask how the 8 hundred can be subtracted. Guide the children to solving the
problem by trading a remaining thousand for 10 hundred,
indicated by the cube crossed out. Subtracting the 1 ten
presents no problem and subtracting the 9 is performed by
bringing a ten to the ones place. Ask a child to construct
the remainder, which is 1748.

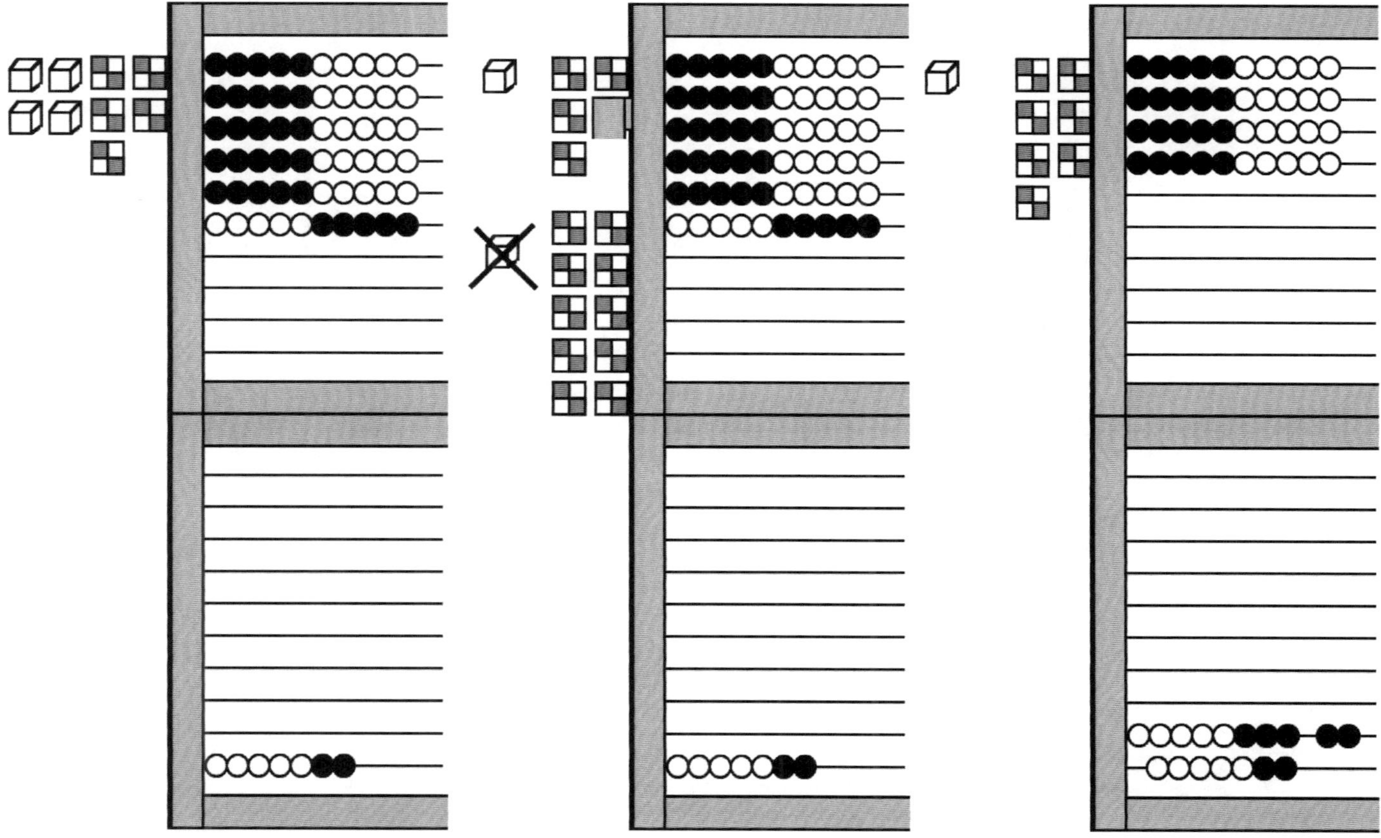

Two other examples for the children to try before working
on their own are: 4823 − 1905 and 5637 − 2498.

4567	3354
− 2819	− 1728
7061	9431
− 3525	− 6670
6604	2651
− 5069	− 1786

BEAD TRADING GAME

This bead trading game and variations are favorites of many children. It concretely lets them experience the "carrying," "exchanging," "regrouping," or "trading" process. There also is a borrowing variation.

Work on the abacus will also become more abstract. Instead of each bead always representing one, a bead can represent a one, ten, hundred, or thousand. The abacus will be used more like the Chinese and Japanese abacuses. With those abacuses, each vertical rod, or wire, represents one position in the place value system. However, to make the calculations easier and more understandable for children, the AL abacus is used with two wires for each position.

The children have been using the abacus in this manner for the ones place. The bead trading games will introduce the ten, hundred, and thousand places with the greatest emphasis on the tens and hundreds.

Basic bead trading

The players each need their own abacuses. Each group of three or four players will also need about eight sets of number cards from 1 to 9. The cards are placed face down on a stack within reach of all the players and may be reused, if necessary, during a game.

The object of the game is to add the quantities as indicated on the cards. A player turns over the top card and enters that many beads in the ones columns. Players do not take turns, but take new cards upon completion of the previous additions. A player who reaches 100 is a champion. A player reaching 1000 is a grand champion. Students have been motivated to reach that 1000 in a 45-minute period!

Whenever the total reaches 10 or more ones, 10 ones are traded for a bead on the ten-wires. That is, ten beads on the one-wires are removed and one bead on the ten-wire is entered. When ten is reached on the ten -wires, they are traded for a bead on the hundred-wires.

Before teaching the game, explain that instead of using a whole row of beads to show a ten, a bead on either of these two wires (point to the 6th and 7th wires) will be the same as a ten. Ask the children to enter 1 ten, then 2 tens; continue to 10 tens. Then ask them to enter quantities 27, 43, and 85. The extra bead can be entered on either side.

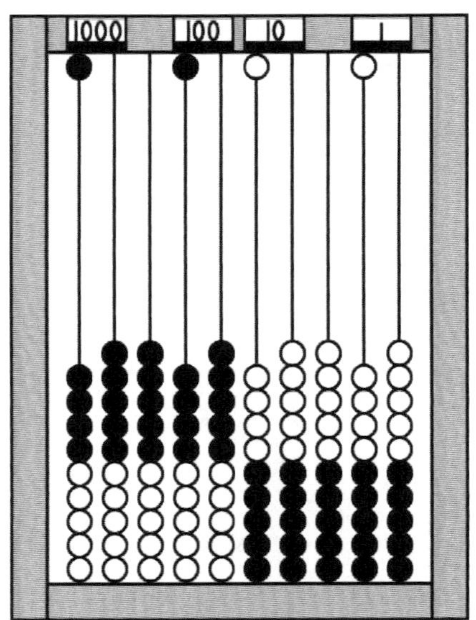

A Chinese abacus.

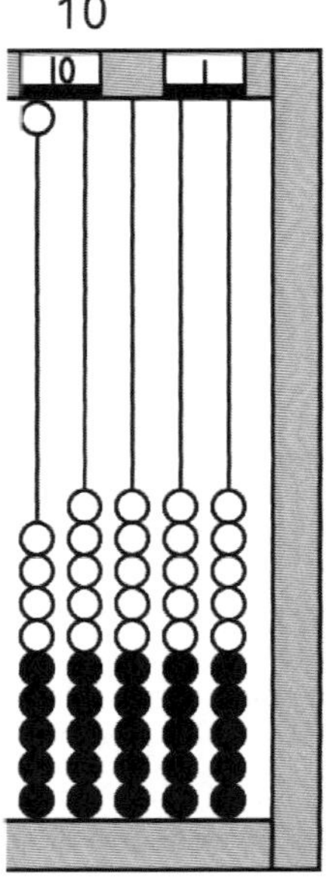

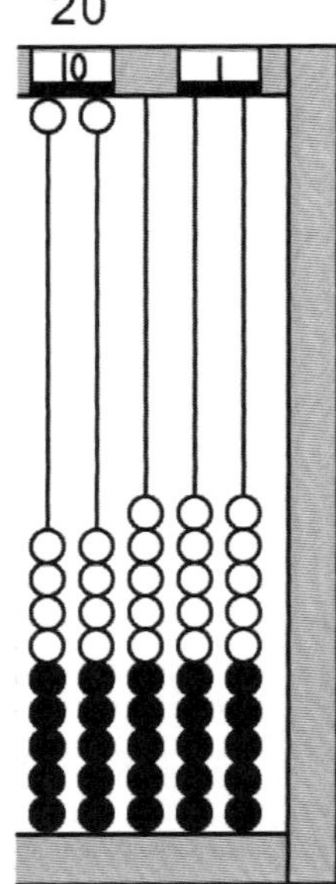

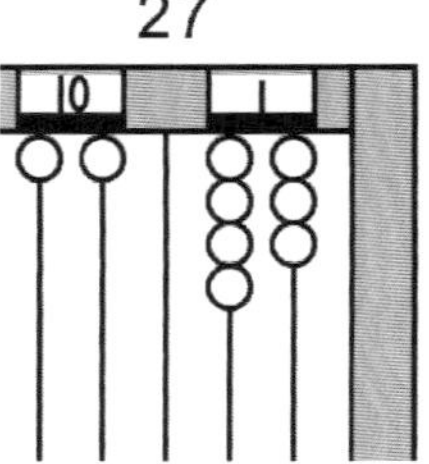

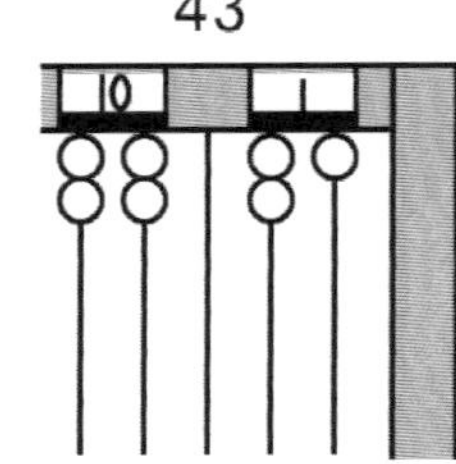

To demonstrate trading, tell the children to enter 8 and then to add 6. Ask, **How much do you have?** [14] **Let's take 10 off and enter it with the tens.** Show trading with the words, "Ready, Set, Trade." The right hand prepared the ten while the left hand prepares the single bead. At the word, "Trade," the hands move in opposite directions. **Now how much do you have?** [14] The children need not completely understand before playing the game.

Demonstrate the procedure with several cards, for example, 3, 9, 6, 4, and 8. Remind the children to keep the two columns as even as possible.

After playing several games, most children will discover shortcuts. For example, to add 8 to 4, they will remove 2 one-beads and add a ten-bead. Or add 9 to 3, they will remove a one-bead and add a ten-bead.

Many children like to continue the Bead Trading over a period of several days. Ask the children to write down their number at the end of the day. Then they can enter and start from it at another time.

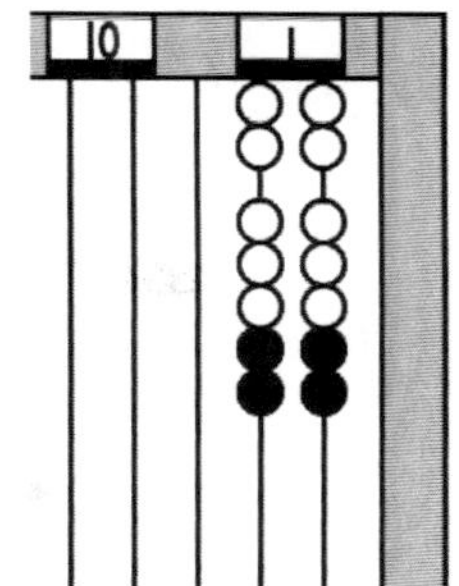
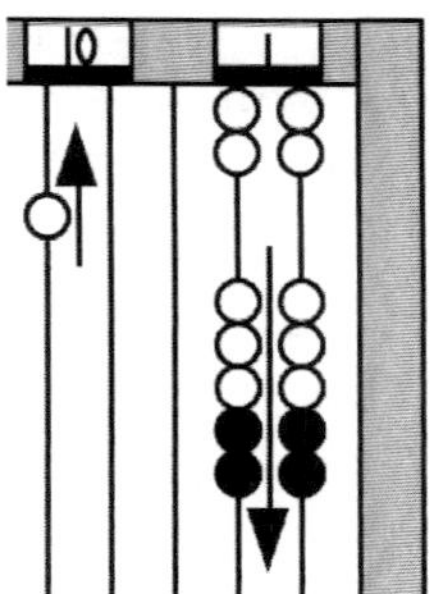
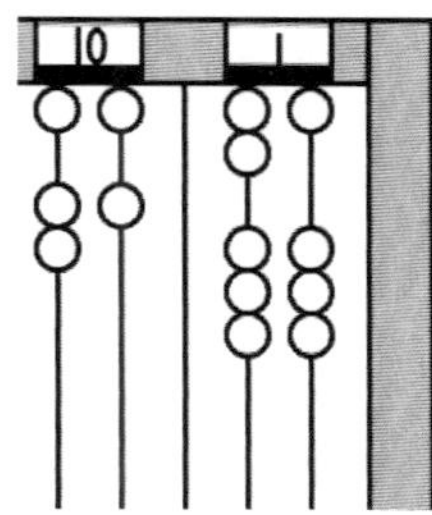

Variations

1. QUANTITIES GREATER THAN 10. Use cards with values between 1 and 100. The product cards from Math Card Games are ideal. If the total entered is 23 and the next card says 36, 3 is entered in the tens columns and 6 is entered in the ones columns. It is preferable that the tens be entered first. The children need to be able to look at the left-most digit and decide its place value.

2. EVEN-NUMBER CARDS. Use only even-numbered cards. Here the children will again experience that even plus even equals even.

ADDITION ALGORITHM

The purpose of this activity is to enhance the children's understanding of carrying which will lead to the usual addition algorithm. With careful attention to the details of adding in this format, many children will discover on their own how to add on paper.

Adding tens without carrying

First, practice with the children entering numbers to 999 in the vertical format.

Start by asking, **Where are the ones places on the abacus?** [the two right wires] Then ask, **Where is the tens place?** [sixth and seventh wires] Ask the children to enter the following numbers:

63   79   34   238   419   5670   1902

Remind them that they may not have more than 9 beads in any place, that is, 10 may not be represented by 10 beads in the ones place.

Draw sketches of or show with an abacus the following combinations and ask them to read and then construct the numbers with the place value cards:

278   639   248   2480   5555   1234   9512

2. Write on the board

$$\begin{array}{r} 45 \\ +\,31 \\ \hline \end{array}$$

Ask the children if they did not sometimes get tired of moving a whole row of beads for one ten. Tell them, **Now they will be using a shortcut for adding, much like the bead trading games.**

Show the children the steps for finding the sum. The first step is to enter 45. Next the 1 is added and the 6 is written in the ones place. Lastly the 3 is added and the 7 recorded. It is important that the partial sums are written as they are formed.

Ask a child to read the sum. Ask if that seems like the right answer. They might want to add it a different way to check.

Give them 23 + 51 and 34 + 25 to add before assigning a worksheet (7-7).

Adding tens with carrying

Tell the children that today's addition work is a little harder than the previous day's work because of the carrying.

Write on the board

$$\begin{array}{r} \bar{2}3 \\ +\,68 \\ \hline \end{array}$$

and place a small line above the 2. Ask them, **What is the first thing you do?** [Enter the 23] **What is next?** [Enter the 8] **Now what special thing must we do before we can write the results?** [Trade 10 ones for 1 ten.] Write the 1. Explain by saying, **We want to show that we have an extra ten, so we will write a little 1 on this line.** Write the 1 directly above the 2.

What do we add now? [The 6] Write the 9. **What is the answer?** [91] Work several more examples with them, such as 58 + 26 and 39 + 47.

ORAL PROBLEMS. A. If the body of a dog measures 38 inches and the tail is 16 inches, how long is the dog when the tail is wagging? [54 inches]

B. Noelle spent 29 minutes reading a book about dogs and 35 minutes reading a book about horses. How much time did she spend reading. [64 minutes]

Worksheets (7-8) could be done in pairs with one child

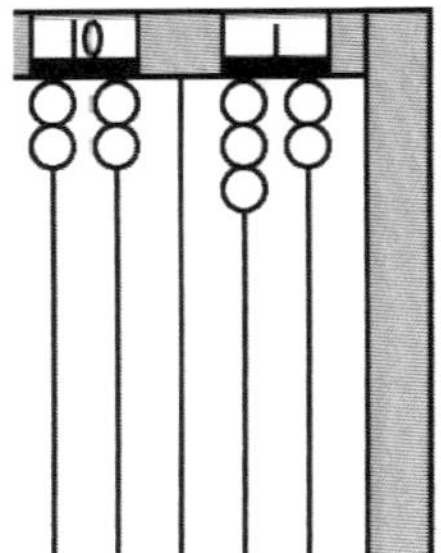 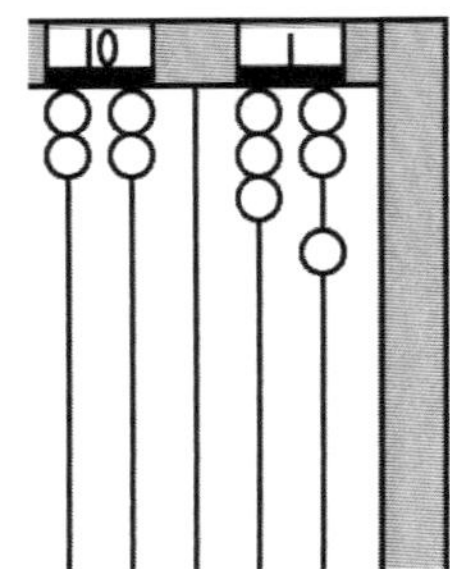
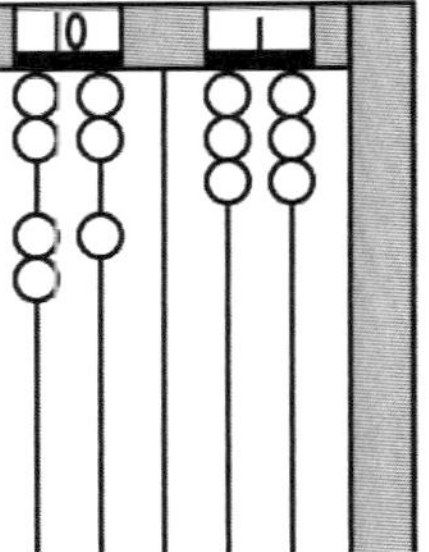 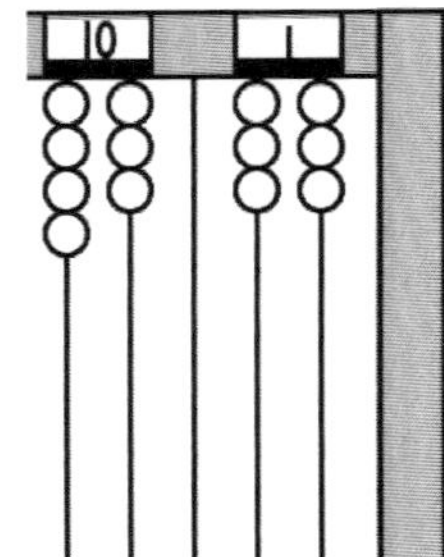

45	26	70	41
+31	+ 3	+ 9	+ 7
32	85	64	73
+ 2	+12	+24	+15
15	63	36	52
+20	+ 4	+42	+34

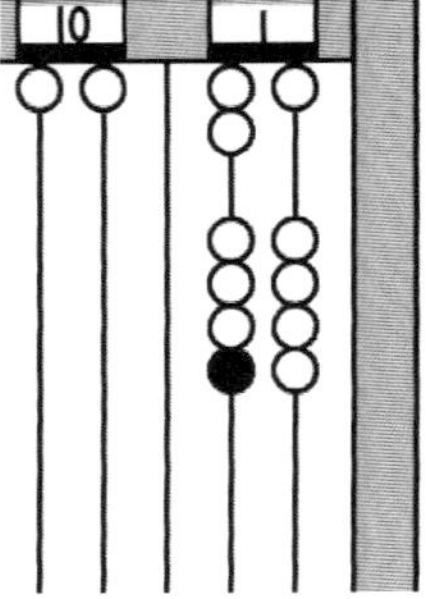 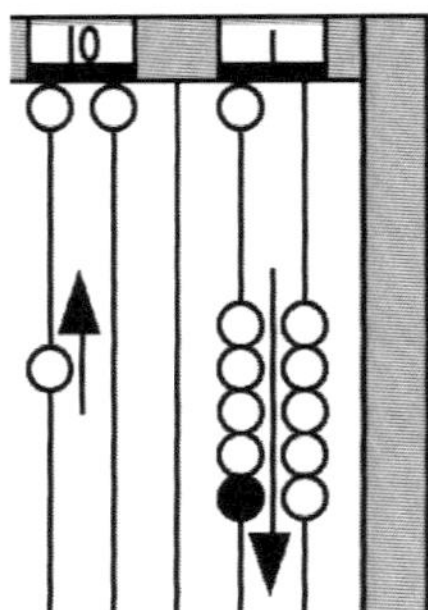
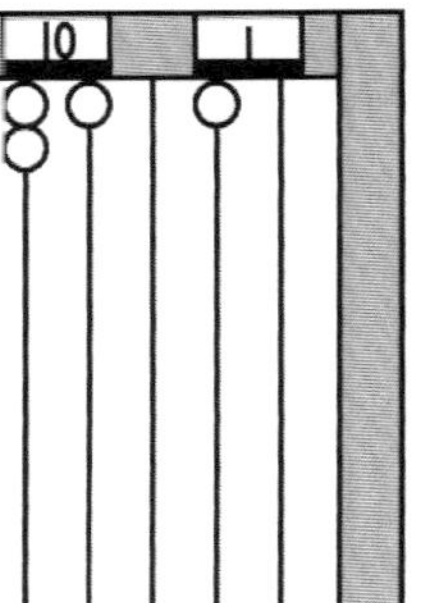 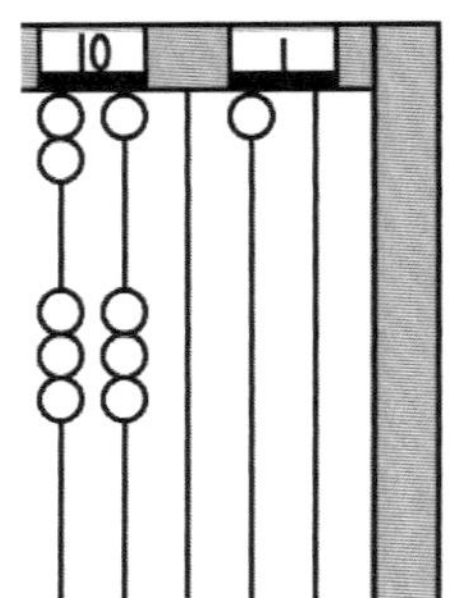

operating the abacus and the other recording the answers. The roles are reversed after each problem.

ORAL PROBLEMS. A. Braden skated for 63 minutes on Saturday and 79 minutes on Sunday. How many minutes did he spend skating that weekend? [142 minutes]

B. Genessa swam 39 minutes on Monday and 43 minutes Tuesday. How long did she swim on the two days? [82 minutes]

Adding hundreds with carrying

1. Write on the board

$$\overline{362}$$
$$+\,275$$

Tell the children that you think they could do that problem with hardly any help. Proceed as before letting the children do most of the work. Also have the group work 793 + 187 and 328 + 403 before assigning them worksheets.

2. Now write on the board

$$\$3.\overline{15}$$
$$+\,.35$$

and call upon a child to read the first amount. [three dollars and 15 cents] Ask another child to read the second amount. [thirty-five cents] The children can add this on the abacus, which adapts very well for money. The dollar amounts are entered on the left side with the light-colored beads on top and the "cents" are entered on the right side with the dark-colored beads on top.

Show the children how to construct a dollar sign by drawing a vertical line through an "S." Be sure they draw the dollar sign and period after each addition.

After the children do the worksheet (7-9 and 7-10), ask them to read the answer to a partner.

ORAL PROBLEM. Amber had saved $2.65 and earned $4.78 shoveling snow. How much money does she have now? [$7.43]

Adding thousands with carrying

Adding in the thousands presents no new techniques. By presenting the thousands earlier in the children's mathematical study, they can see how the addition algorithm actually works. Challenge them with worksheets without instruction. Encourage a child with a question to ask a classmate.

Some of the children will have discovered how to add without the abacus. For the others, either individually or as a group, write

23	48	57	42
+68	+45	+19	+39

52	37	15	24
+26	+34	+17	+28

69	38	28	59
+30	+28	+47	+37

362	574	349
+275	+189	+608

892	476	213
+97	+374	+548

495	398	657
+ 5	+453	+184

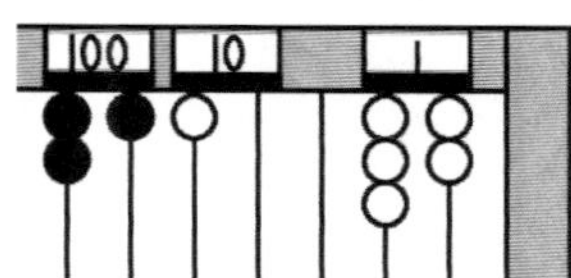

$3.15	$ .25	$1.98
+ .35	+ .25	+ .09

$6.89	$ .75	$2.63
+ .61	+ .75	+ 1.37

$4.99	$6.09	$3.72
+ .12	+ 2.65	+ 4.68

$$\overline{\overline{2579}} \\ +\,3684$$

and ask them to write the answer.

$$\underline{111} \\ 2579 \\ +\,3184 \\ \overline{5763}$$

Now ask the following questions: **How much is 9 + 4?** [13] **We write the 3 here and the 1 up here.** Point to the 3 below the line and the 1 above the 7. **How much is 1 + 7 [8] + 8?** [16] **We write the 6 here and the 1 up here.** Point to the appropriate places. **What do we do next?** [Add 1 + 5 + 1] Finish with the 2 + 3.

Use previously worked examples for further help, if needed. Give the children a worksheet (7-11 and 7-12) with more examples.

3687	8165
+ 2147	+ 943
6083	4759
+ 1892	+ 2643
8840	2755
+ 659	+ 1895

Adding three numbers

In order for a column of numbers to total 20 or more, or the number carried to be other than 1, three or more numbers must be added together. But first approach adding three numbers where the columns total less than 20.

Write the following three numbers on the board:

$$134 \\ 28 \\ \underline{+\,51}$$

Ask the children to add the ones [13] and to record the 3 and 1 in the appropriate places. Next ask them to add the tens and hundreds in the same way.

Next, write on the board:

$$278 \\ 369 \\ \underline{+\,183}$$

Ask the children to start the same way, by adding the ones. [20] Challenge them to think about what to do with the 2 tens. When they understand how to proceed, give them the appropriate worksheet (7-13 and 7-14).

ORAL PROBLEMS. A. In Lisa's school, there are 3 classes with her grade. How many students are there if they have 21, 19, and 18 students? [58 students]

B. Park Elementary has 58 teachers. West Elementary has 36 teachers and North has 47 teachers. How many teachers do the three schools have? [141 teachers]

C. Shannon's school has 386 students. The other two schools in her city have 1,259 and 797 students. How many students does her city have? [2442 students]

134	153	267	536
28	60	324	247
+ 51	+ 7	+ 14	+ 53
807	267	245	167
99	308	68	358
+ 26	+ 19	+ 97	+415

SIMPLIFIED SUBTRACTION

Of the five different algorithms for subtraction, simplified subtraction is the simplest to learn and easiest to perform. Although simplified subtraction will be the method shown in this book, the traditional method of subtraction works as well on the abacus.

Subtracting tens without borrowing

Subtraction, like division, proceeds from left to right. Write on the board

$$\begin{array}{r} 76 \\ -\,34 \\ \hline \end{array}$$

and tell the children, **This is subtraction, so we will start next to the minus sign.** Ask them to enter 76, subtract 3 tens and write 4, and finally subtract 4 ones and write 2. Give them a few examples like 89 – 25 and 48 – 16 before handing them a worksheet (7-15).

2. Ask the children what would be the sum if the remainder 21 were added to the subtrahend 34. If they are unsure, ask them to add them.

To help them develop the habit of checking their subtraction, give them two strips of paper. One is used to cover the minuend and the other is placed below the remainder. Ask the children to add the two numbers they see, then remove the upper strip, and compare.

Subtracting tens with borrowing

Write on the board

$$\begin{array}{r} 4\,^-3 \\ -\,2\,5 \\ \hline \end{array}$$

with a small line before the 3. Ask the children to enter the 43 and say, **Can you subtract 2 from 4?** [yes] **But before we write down the 2, look ahead; can we subtract 5 from 3?** [no] **What must be done?** Guide the children to trade a ten for 10 ones and say, **Now let's write down what we have. How many tens are there?** [1] **How many ones?** [13] Explain that the 3 can be changed into 13 by writing a small 1 on the little line. Continue with subtracting 5 from 13 and writing the result.

Go through the same process with 57 – 29 and 30 – 12. Assign a similar worksheet (7-16). Include the little lines initially on the worksheets, which help the children know where to write the superscripts. These will also be needed for division.

ORAL PROBLEMS. A. Natalie is 7 years old; her mother is 28 years old. What is the difference in their ages? [21 years]

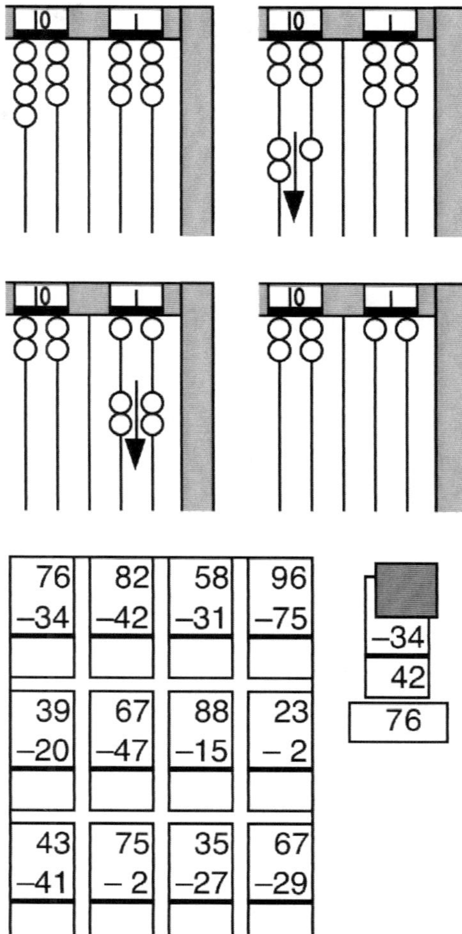

76	82	58	96
−34	−42	−31	−75

39	67	88	23
−20	−47	−15	− 2

43	75	35	67
−41	− 2	−27	−29

−34
42

76

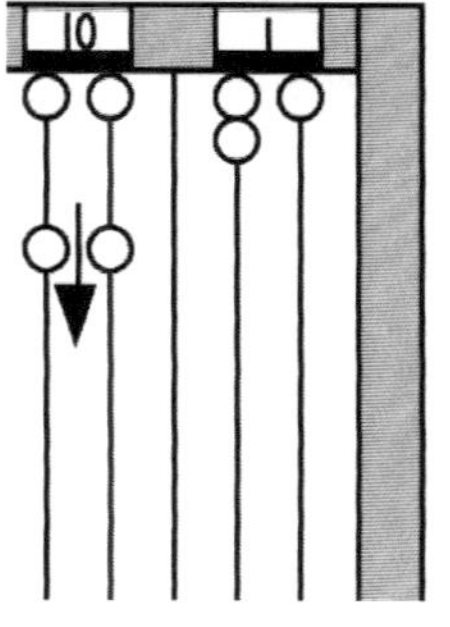
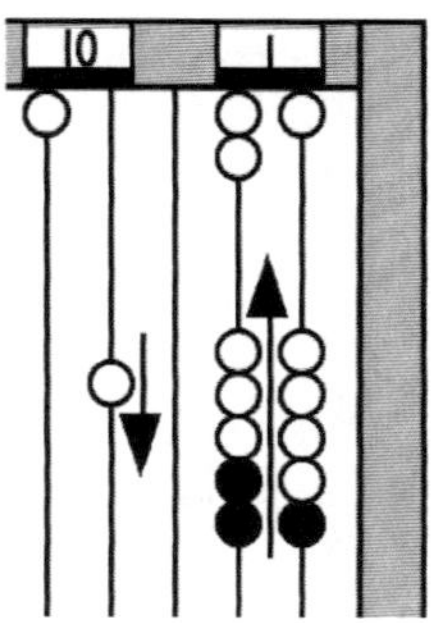
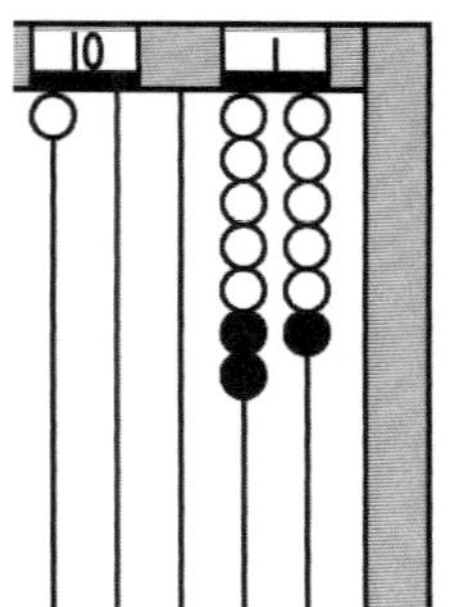
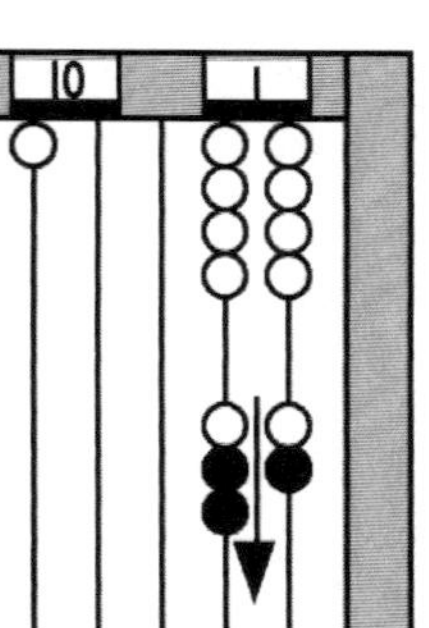

When the children arrive at an answer, remind them to think if it makes sense. If their answer is 11, ask them if that could be possible.

B. Kristen's father is 41 and her brother is 15. What is the difference in their ages? [26 years]

C. Danny is 6 years older than his cousin who is 17. How old is Danny? [23 years]

Bead trading with subtraction

Start with a score of 99 for the first game. Champions are those reaching 0. When subtracting numbers greater than 10 such as those on the product cards, it is important that the tens be subtracted first because simplified subtraction works from left to right and it is easier to do mentally.

Subtracting hundreds with borrowing

1. Subtracting in the hundreds has no new skills to master, except perhaps a closer look to determine whether trading is necessary. Write on the board

$$\begin{array}{r} 863 \\ -481 \\ \hline \end{array}$$

and ask, **Where do we start.** [8 – 4] Ask, **Before we write down 4, we must look: is 63 more than 81?** [no] Go through the trading process and write down the 3 in the hundreds place and the 1 next to the 6.

Ask, **What do we subtract now?** [16 – 8] **Before we write it down, what do we have to check?** [Is 3 more than 1] **So, what do we write?** [8] Continue with 3 – 1.

Before giving the children individual problems (7-17), work the examples 703 – 258 and 627 – 118.

ORAL PROBLEMS. A. A baby boy weighs about 7 pounds at birth. A grown man weighs about 150 pounds. How much less does a baby weigh than a man? [143 pounds]

B. At birth, a girl measures about 53 cm long. An adult woman is about 162 cm tall. How many cm must the baby grow to become a woman? [109 cm]

C. A newborn blue whale is about 23 feet long. An adult whale may be 95 feet long. How much will the baby whale grow? [72 feet]

D. Twenty-three feet, the length of the baby whale, is equal to 701 cm. A human baby is 53 cm long. How much longer is the newborn whale? [648 cm]

E. A large dinosaur could be 79 feet long. How much shorter is this than the blue whale, which is 95 feet long? [16 feet]

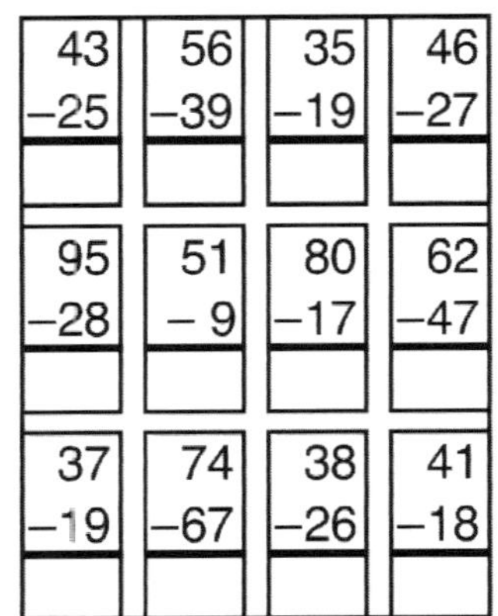

43	56	35	46
−25	−39	−19	−27
95	51	80	62
−28	− 9	−17	−47
37	74	38	41
−19	−67	−26	−18

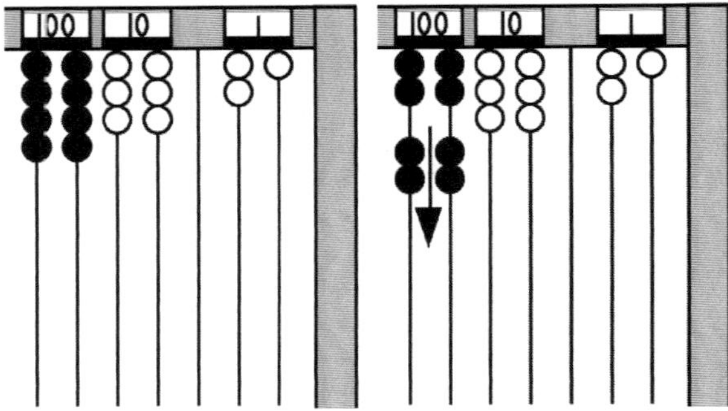

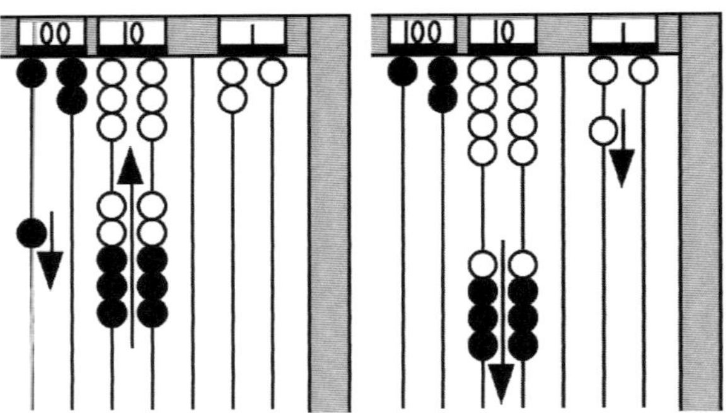

863	939	432
− 481	− 246	− 119
884	115	940
− 347	− 94	− 573
807	397	233
− 615	− 289	− 217

2. Write on the board

$$
\begin{array}{r}
592 \\
-294 \\
\hline
\end{array}
$$

and note the interesting problem arising because of the two 9s. Start as usual with $5 - 2 = 3$, but before the 3 is written, check and notice that 92 is not more 94, so trading is necessary. Therefore, 2 is written and 9 is subtracted from 19 leaving 10. However, 10 is reduced to 9 because 2 is not greater than 4. See the figures.

Ask the children to work on $753 - 656$ and $204 - 103$ and assign an appropriate worksheet (7-18).

3. The transition to subtracting exclusively on paper will be possible for many children now. The algorithm is as follows: subtract the leftmost digits. If the minuend to the right is more than the subtrahend to the right, write the remainder and continue by subtracting the next digits on the right. If the minuend is less, write 1 less than the remainder and add 10 to the next minuend digit on the right. Continue by subtracting the number directly below it.

For variety, worksheets (7-19 and 7-20) with money problems can also be given at this point.

ORAL PROBLEMS. With the following information, let the children make up and solve problems.

A. Aaron saved $3.82 and Scott saved $8.79. [They could either find the total saved, $12.61, or the difference saved, $4.97.]

B. The city of Greenville has a population of 695 and the city of Redburg 893. [Total population is 1588; difference is 198.]

C. During the month of June, children checked out 738 books from the library. In July, the children checked out 593. [Total 1331; difference 145.]

D. A pair of used skates cost $29.99 and a pair of used skis costs $63.98. [Both $93.97; difference $33.99]

Teach the children the following trick that they can then use with family and friends: Choose any number more than 10. Subtract 5. Double the results. Subtract the number you chose. Then say, "Tell me your last number and I will tell you what number you started with." To the number given add 10 to find the original number.

For example, if 25 were chosen, the remainder after subtracting 5 is 20. That doubled is 40 and $40 - 25 = 15$. The number 15 is given to the other person who adds 10 and arrives at 25, the original number.

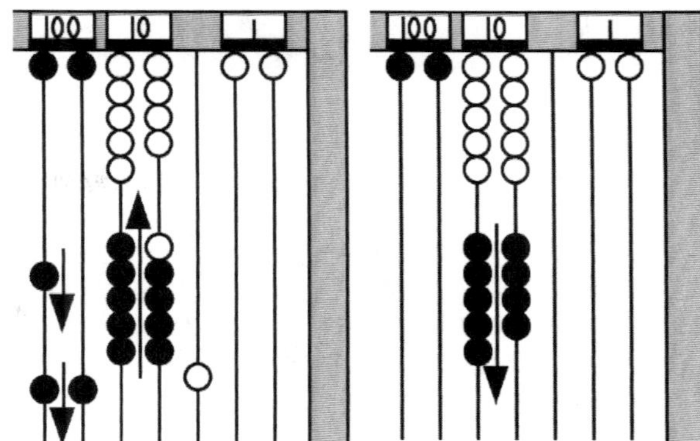

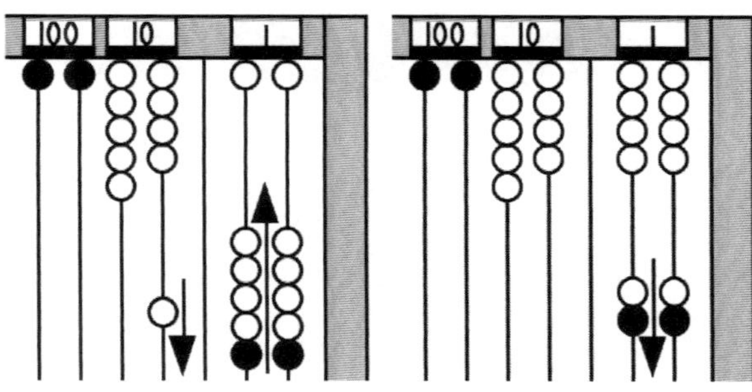

592	436	658
− 294	− 238	− 555
884	407	701
− 385	− 305	− 107
396	754	514
− 296	− 258	− 316

$3.68	$7.05	$9.00
− 2.49	− 4.85	− 1.75
$6.47	$7.50	$5.33
− 2.38	− 3.75	− 4.74
$4.00	$8.61	$2.12
− 2.12	− 8.59	− 1.99

Subtracting thousands with borrowing

Write on the board

$$\begin{array}{r} 8362 \\ -\ 2854 \end{array}$$

and challenge the children to see if they can subtract it without help. There is no reason for the children to continue writing the extra little 1s, the superscripts. In the above example, if 8 cannot be subtracted from 3, then think what 8 is from 13.

The work on the abacus is shown on the right. The thinking without the abacus is $8 - 2 = 6$. But $3 < 8$ so write 5. Next $13 - 8 = 5$ and $6 > 5$ so write 5. Then $6 - 5 = 1$, but $2 < 4$ so write 0. Lastly $12 - 4 = 8$, which is written as 8.

Give the children work on worksheets (7-21 and 7-22).

ORAL PROBLEMS. A. Highview School has 387 students and Bellaire School has 566 students. How many students do the two schools have? [953 students]

B. A town has 3292 students in elementary and 1488 in junior high. How many students is that altogether? [4780 students]

C. The town of Silver Lake has 505 residents. If 269 are female, how many are male? [236 residents]

D. Neil has $10.00. He wants to buy 2 items, one costing $4.39 and the other $5.25. Does he have enough money? [yes]

E. For the first performance of a concert, 5329 persons attended. For the second performance, 3872 persons attended. How many attended both concerts? [9201 persons]

F. Carmen traveled 270 miles on Monday, 348 miles on Tuesday, and 413 on Wednesday. How many miles did he travel? [1031 miles]

G. Rhett helped buy plants. Tomatoes cost $1.79; peppers cost $2.69; and melon costs $.89. What is the total cost? [$5.37]

H. Elizabeth wanted to buy marigolds for a garden. The marigolds cost $3.69. She had only $2.50. How much more did she need? [$1.19]

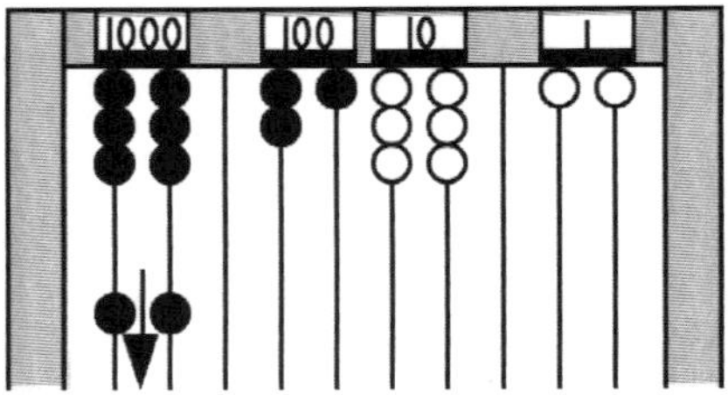
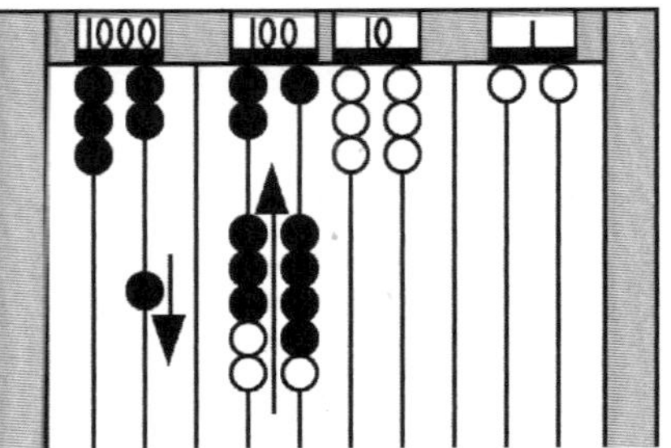
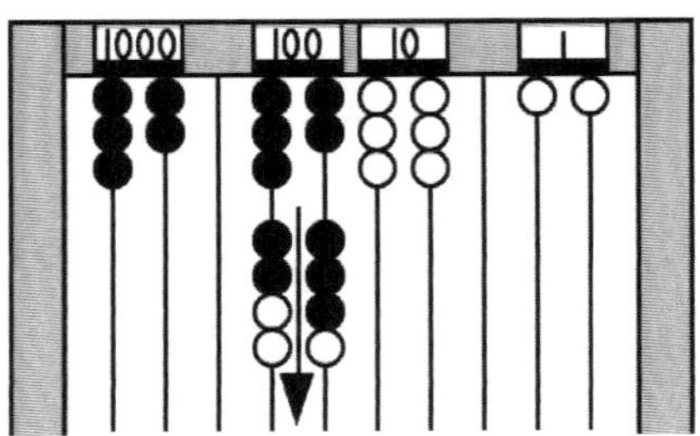
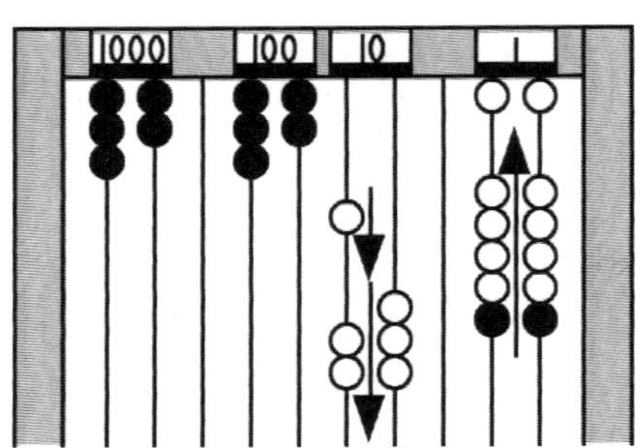
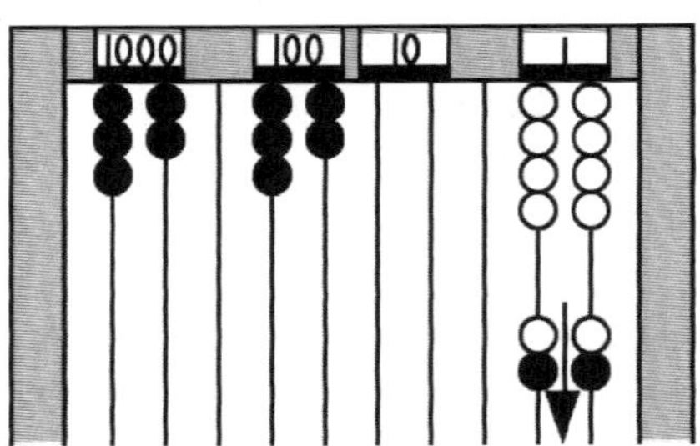

8362	3720
− 2854	− 1919
6524	3863
- 2476	- 2983
2033	5813
− 949	− 2254

Unit 8
Multiplication

Multiplication is introduced through arrays, an essential model for finding areas, and through repeated addition. The facts are approached through skip counting. Skip counting will be used later working with multiples, especially the least common multiple (LCM) and the greatest common factor (GCF), both encountered in fractions.

While the commutative and associative laws are discussed, the distributive law is not mentioned in this unit. The concept of the distributive law is easy to understand with an abacus, but very difficult to write on paper because of parentheses and the length of the equation. The parenthesis will be discussed in the unit on Other Topics.

This unit continues with multiplying two-, three-, and four-digit numbers by a single-digit number. It concludes with multiplying by two-digit numbers.

SKIP COUNTING

Skip counting was introduced earlier with the 2s, 5s, and 10s. The 3s, 4s, 6s, 7s, 8s, and 9s will be studied now.

On the abacus

Tell the children that they are going to work on skip counting. Draw on the board an array of 5 by 2 squares as shown. Review by asking the children to enter the 2s up to 20 on the abacus. Then call on different children to write the quantities in the squares.

For the 3s, only three quantities are written in a row. On the abacus any extra beads are kept to the right.

Continue with the remaining quantities. For the 4s, 6s, and 8s, use the 5 by 2 grid; for the 7s, use the 3 grid.

The 9s use the 5 by 2 array, but the second line of the 9s is written from right to left to enable the children to discover the digits reversing. With the numbers in these arrangements, various strategies for memorization become possible.

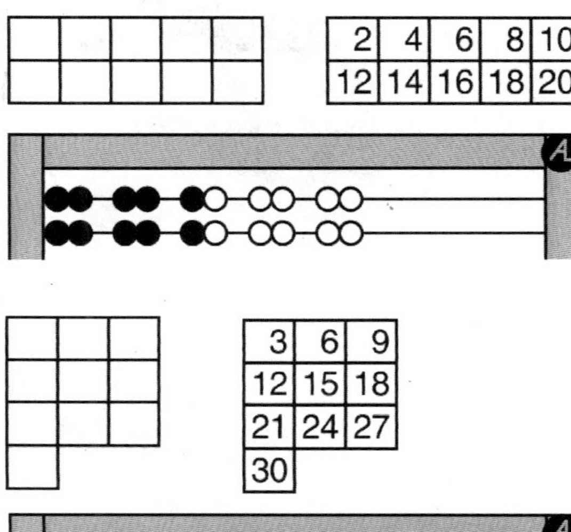

Give the children a worksheet (8-1) for working independently and for future reference. Tell them that we do want them to learn these in the near future.

FUTURE REVIEW. With the completed worksheet before them, ask the children these and similar questions: What is three 3s? Two 4s? Five 4s? Ten 6s? Six 2s? They only need to count over that many blocks (or groups on the abacus) to find the answer. For example, two 4s is found in the second block of the 4s.

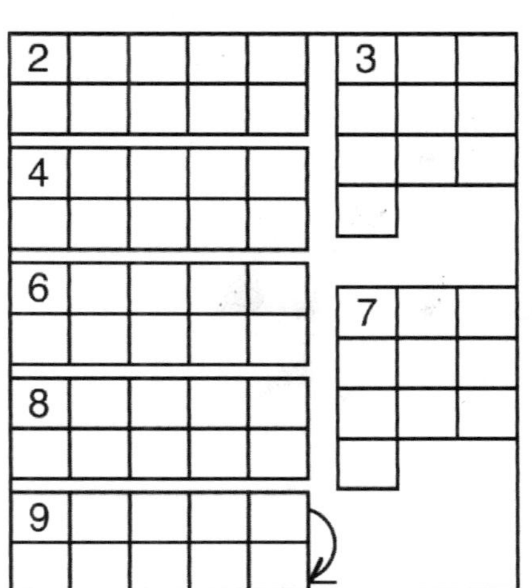

Patterns on the hundred chart

Some very interesting patterns and relationships can be made on the hundred chart (worksheet 8-2). One way to see these is with transparencies. Prepare nine charts on transparencies or tracing paper. Distribute them to nine children for coloring with non-permanent pens. One child colors the 2s, another the 3s, and so forth. Ask the children to describe the various patterns. The 3s are shown in the figure.

To compare the relationships, place, for example, the 2s on top of the 3s, and ask the children to see which numbers are included in both the 2s and 3s. Try also the 4s and 8s or the 7s and 9s.

For individual work, give them copies of the hundred chart (8-2) to color or mark for various combinations.

Strategies for memorizing

To help the children memorize the skip counting patterns, give them the following pointers.

The 2s are arranged in two rows that show the ones repeating. The second row is 10 plus the first row.

2	4	6	8	10
12	14	16	18	20

With the 3s arranged in groups of 3, consider the tens place in each row: the first row is less than ten, the second row is in the teens, and the third row is in the twenties. Also observe that the digit in the ones place increases by one, starting with 0 at the lower left and moving up and continuing at the bottom of the next column.

3	6	9
12	15	18
21	24	27
30		

The 4s are grouped in two rows with the second row 20 more than the first row. Also note that the first row of 4s is the 2s with every other number removed.

4	8	12	16	20
24	28	32	36	40

The 5s are easily learned. Children like to recite them in singsong fashion.

The 6s can be thought of as the even 3s for the first row. In the second row the ones are the same and the tens are 3 more; that is, it is 30 more than the first row.

5	10
15	20
25	30
35	40
45	50

6	12	18	24	30
36	42	48	54	60

Even the 7s have a certain pattern. The ones are each used once, as with the 3s and the 9s. In each column, starting at the right, the ones increase by one, while in the rows, the tens increase by 1.

The 8s are every other 4 for the first row. Also notice that in each row, the ones are the even numbers in descending order. In each row the tens increase by 1 and the second row is 40 more than the first row.

7	14	21
28	35	42
49	56	63
70		

8	16	24	32	40
48	56	64	72	80

The 9s are the most interesting of all; the sums of the digits in each case equals 9. The digits can also be reversed to obtain another multiple as shown in the second row. Also note in the rows that the ones decrease by 1 as the tens increase by 1.

9	18	27	36	45
90	81	72	63	54

SKIP COUNTING CONCENTRATION. Children immensely enjoy playing this simple game, which teaches the skip counting patterns. Prepare sets of cards for each pattern (or use the product cards from the Math Cards deck) and put each set in a separate envelope, which is marked with the skip counting patterns described above. The players may refer to the envelopes as necessary.

This is an in-order concentration for two players. The players each choose an envelope and mix the two sets together and lay them face down in rows of five cards each.

The object is to be the first player to pick up, *in order*, her complete set of product cards. The first player turns over a card so both can see it. If it is the card needed, she picks it up and receives another turn. If it is not the card needed, it is returned face down to the original place. The other player then takes a turn. Players continue to take turns until one of them has picked up all her cards.

Both players will at times need the same number, which adds interest and excitement to the game.

To practice memorizing the patterns, the children can also play the four simple card games given in Unit 1. After this type of work give them worksheets (8-3 and 8-4), which has some numbers missing.

Review of adding and subtracting

An important part of computation is aligning the numbers prior to hand computation. Write on the board

$$1145 + 706$$

Ask the children how they would add them on paper. If they need help, suggest they write the first number. Next they decide the place value of the first digit of the second number and write it directly below the corresponding digit of the first number. The other digits are written in order. Give them a worksheet (8-5) with this type of problem.

INTRODUCING MULTIPLICATION

Multiplying is introduced as repeated addition but in the context of an array; that is, the same number is "taken" so many times. Memorizing the facts is not of concern at this point, but memorizing the skip counting patterns should continue.

Basic multiplying

Tell the children that they are going to start multiplication. With the wires horizontal ask a child to enter 5. Ask another child to enter 5 on the next wire. Repeat a third time. Summarize, **Now we have 5 taken three times. This is how we write it:**

$$5 \times 3 = \_\_$$

3		9
1_	15	_8
21	_4	2_
_0		

7	1_	21
2_	_5	4_
_9	56	63
70		

4	8	12	16	20
2_	2_	32	3_	_0

6	12	1_	24	__
3_	42	48	5_	6_

8	1_	24	3_	40
48	_6	__	_2	80

9	1_	_7	3_	45
90	81	_2	6_	_4

$$\begin{array}{r} 1145 \\ + 7 \end{array} \qquad \begin{array}{r} 1145 \\ + 706 \end{array}$$

1145 + 706
1580 + 424
2460 + 4189
3745 - 81
8504 + 971
4027 - 442
74 + 6307
2179 + 513
3822 - 1585
7511 + 429
2312 + 316
5731 + 703

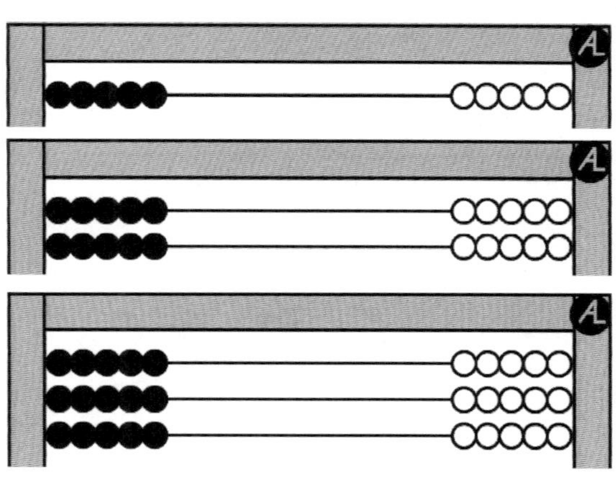

(The term "times" will be introduced later.) Ask, **How much is it?** [15] Ask a child to write the answer.

Write on the board $6 \times 4 =$ ___. Ask the children to read it [6 taken 4 times], then to enter it on their abacuses, and to find the answer. Then ask to enter 7×8. Encourage them to see the tens groupings as shown.

Next, ask them to find 9×5. Since they know 10×5, subtracting 5 from 50 will give the product of 45. This also works for multiplying by 8s.

Give them a page (8-6, 8-7) of problems, letting them find the answers in any manner they choose.

FUTURE REVIEW. To help the children see these facts visually, enter an array and show it to the children for 2-3 seconds. Then ask them to state the multiplication fact and the product. (For the classroom, an overhead abacus is available from the publisher.)

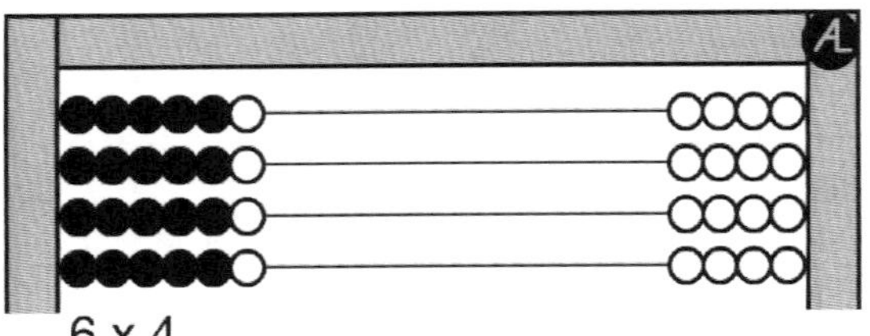
6 x 4

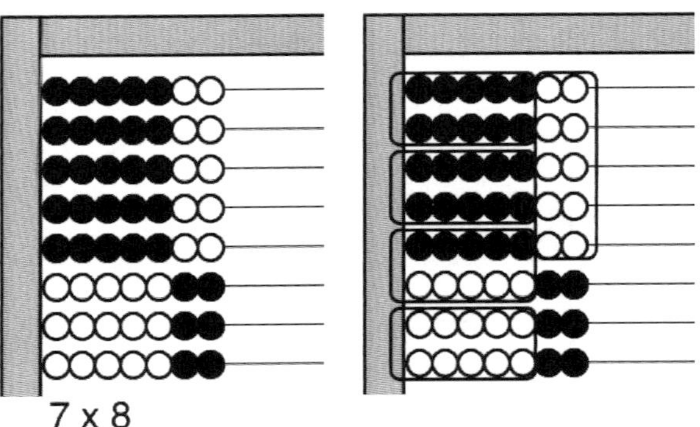
7 x 8

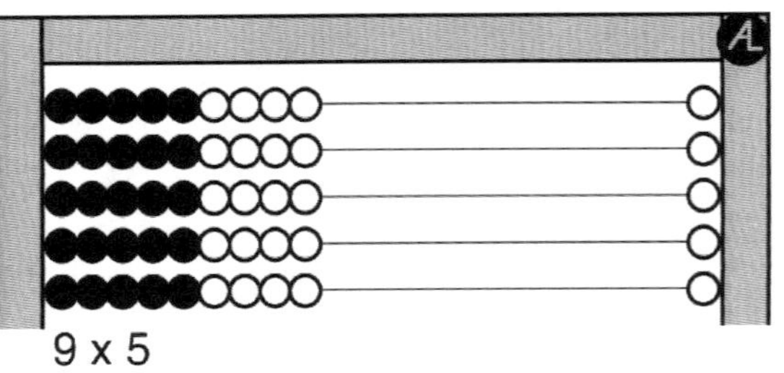
9 x 5

Multiplying with 0s and 1s

1. The object of this lesson is that the children discover and learn the rules associated with multiplying with zeroes and ones. Work with them only long enough to be sure they understand how to do it. Ask them to look for patterns.

Write on the board

$$0 \times 3 = \underline{}$$

and ask, **How much do we enter each time?** [0] **How many times do we enter it?** [3] Go through the motions of entering nothing on 3 wires. **So what is the answer?** [0]

Next write on the board

$$8 \times 0 = \underline{}$$

Give the children time to think how they would do it before showing them. **How much do we enter each time?** [8] **How many times do we enter it?** [0] Move 8 beads on the top wire a slight distance, stop, and say, **We cannot enter it even once. So how much is 8×0?** [0]

If necessary give them more examples, 0×4, 9×0. Then give them a worksheet (8-8) with 0s.

FUTURE REVIEW. After the "0" worksheet is completed, ask what happens when there is a 0 in a multiplication problem. [The answer is 0.]

2. Tell the children that now they will work with multiplying 1s. There is no confusion with problems like 1×7, but 7×1 can cause difficulties. Write on the board

$$7 \times 1 =$$

9 x 3 = ___	0 x 3 = ___	7 x 1 = ___
7 x 5 = ___	9 x 0 = ___	1 x 5 = ___
5 x 3 = ___	5 x 0 = ___	8 x 1 = ___
8 x 8 = ___	0 x 7 = ___	1 x 3 = ___
6 x 9 = ___	4 x 0 = ___	1 x 7 = ___
6 x 3 = ___	0 x 10 = ___	9 x 1 = ___
9 x 10 = ___	6 x 0 = ___	1 x 9 = ___
9 x 4 = ___	0 x 2 = ___	5 x 1 = ___
5 x 9 = ___	0 x 8 = ___	4 x 1 = ___
8 x 5 = ___	7 x 0 = ___	0 x 1 = ___
9 x 9 = ___	0 x 6 = ___	1 x 2 = ___
5 x 6 = ___	0 x 0 = ___	10 x 1 = ___
9 x 5 = ___	1 x 0 = ___	1 x 1 = ___
8 x 6 = ___	0 x 9 = ___	1 x 4 = ___

Ask, **How many times do we enter the 7?** [1] Ask them to find the answer. Assign the worksheet (8-8).

FUTURE REVIEW. After the work with the "1" worksheet is completed, ask, **What happens when you multiply by 1?** [The answer is the number being multiplied by 1.]

Multiplying with take and give

Review the names given to answers. Ask, **What do we call the answer in addition?** [sum] **What do we call the answer in subtraction?** [remainder or difference] **The answer in multiplication is called the PRODUCT. In the problem of 3 taken 2 times equals 6, what is the 6 called?** [product]

Another way to find the product is called "take and give." Younger children often enjoy working with the take and give that they ask for more of the "harder" problems.

Tell the children that you are going to show them a short-cut for solving problems with higher numbers. Write on the board

$$9 \times 3 = \underline{\quad}$$

and ask them to enter the problem. Explain that the product can be read easily if the rows are tens. Therefore, to make the first row a ten, TAKE a bead away from the last row and GIVE it to the first row. Do the same thing for the second row. Then the product can be read as 27.

Ask them to work problems, 7×8 and 8×9. Any number of beads may be removed at a time as long as the same number is restored on another wire. Provide worksheets (8-9, 8-10).

Also give them worksheets (8-11) with the problems written in the vertical form; tell the children that they are worked the same way. Also tell them that there is another way to read a multiplication problem; 9×3 can be read as "9 times 3."

Writing multiplication tables

1. Say to the group that they will be writing the multiplication tables. Although there are no new facts, the tables organize all the multiplication facts, which they need to learn. They already know some of them. After the tables (8-12 to 16) are completed, they can be fastened together.

Write on the board a sample table, for example, the 2s as shown on the mini worksheet shown on the right.

To start, enter the first 2 and record the product. Subsequently add another 2 and write the result. Do not clear until the entire table is completed.

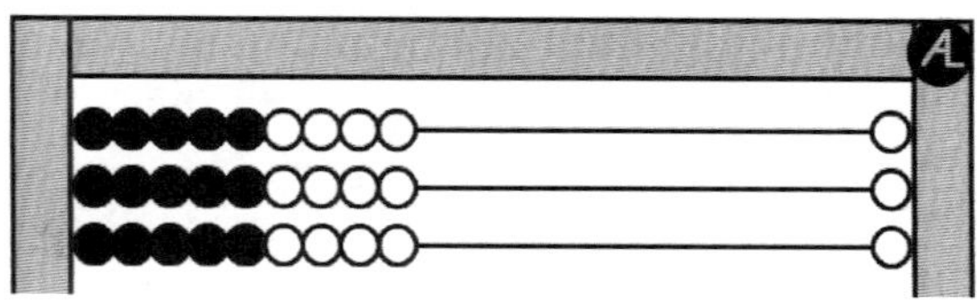

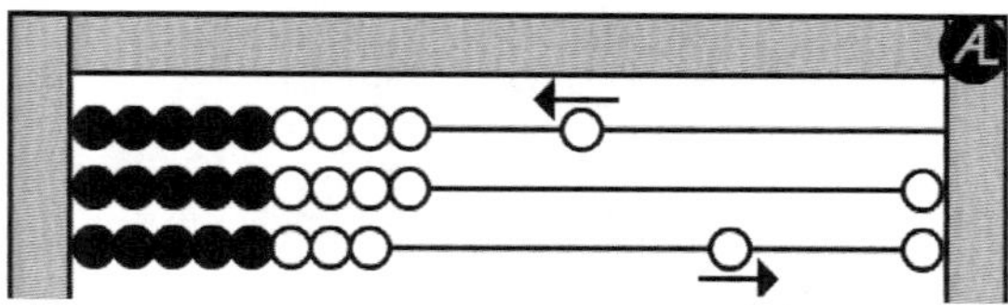

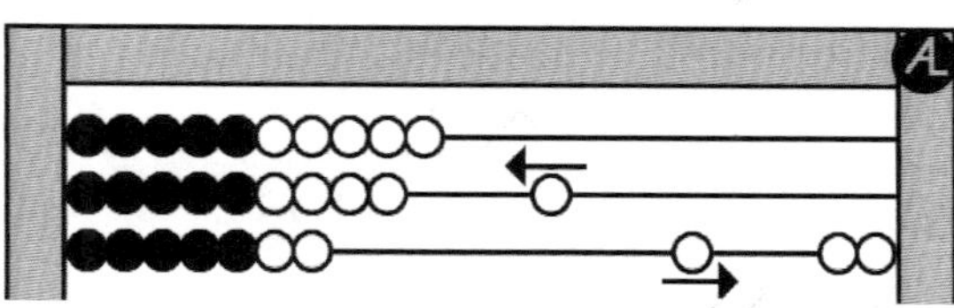

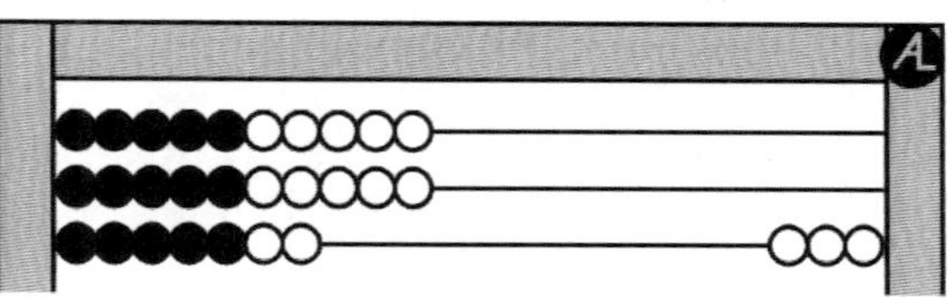

9 x 3 = __	6	9	3
7 x 5 = __	x 5	x 3	10
5 x 3 = __			
8 x 8 = __	8	6	6
6 x 9 = __	x 4	x 3	x 6
6 x 3 = __			
9 x 10 = __	5	2	7
9 x 4 = __	x 9	x 1	x7
5 x 9 = __			
8 x 5 = __	4	7	8
9 x 9 = __	x 1	x 3	x 2
5 x 6 = __			
9 x 6 = __			
8 x 6 = __			

2 x 1 = __
2 x 2 = __
2 x 3 = __
2 x 4 = __
2 x 5 = __
2 x 6 = __
2 x 7 = __
2 x 8 = __
2 x 9 = __
2 x 10 = __

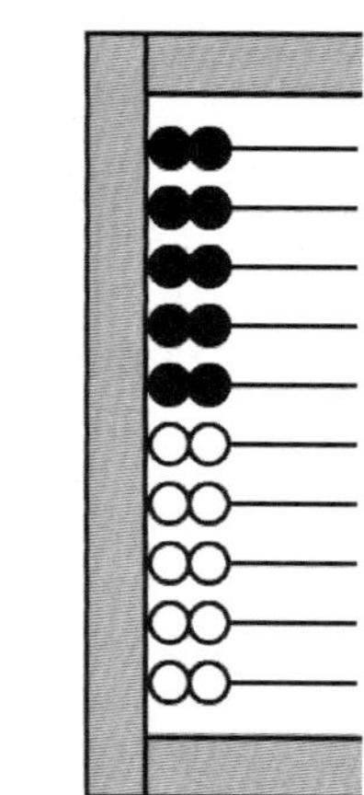

2. When the small multiplication tables are completed, the children can build the large multiplication table by copying from the small tables. Give them an 11 by 11 grid (8-17) for this purpose. Ask them to do a careful job because the table will be used in the future.

Using the table

To further help the children realize that multiplication is repeated adding, this lesson asks them to write the equivalent multiplication fact and to use the multiplication table to find the answer.

Write on the board

$$4 + 4 + 4 = \underline{\quad}$$

Ask if they could write that equation another way. [4 × 3 = _] Then show them how to use the table. Locate the first number along the top row and count down the number of rows equal to the second number. The answer is 12. Also ask them to show it on their abacuses.

Give them a worksheet (8-18) as shown. Also give them a similar worksheet (8-19) with the additions in columns.

ORAL PROBLEMS. A. There are 9 players on a softball team. How many players are on 2 teams? [18 players]

B. A school has 5 teams with 6 on a team. How many students are on the teams? [30 members]

C. Ten people are sitting on chairs in a circle. How many legs are on the floor? [either 20, or 40, or 60 legs depending upon which legs are counted, the people's or the chairs' or both]

D. Lindsay is waiting 3 weeks for a vacation to start. How many days is that? (If the children do not know how many days in a week, guide them to using a calendar to find out.) [21 days]

Guess the sign

Until now, a single number is all that the children have seen following the equal sign. Eventually they will need more complex equations like 25 = (2 × 5) + (3 × 5).

This activity will introduce them to the equal sign in a new place. They will also be asked to determine what sign (+, -, x, or =) will make the equation true. One way this can be done is to guess and check until the guess is right. Guessing is a valid strategy for problem solving. Of course, careful observation improves the chances for a good guess.

Write on the board

$$6 + 7 = 13$$

and say, **If 6 plus 7 equals 13, we can also say that 13 equals 6 plus 7.** Write on the board

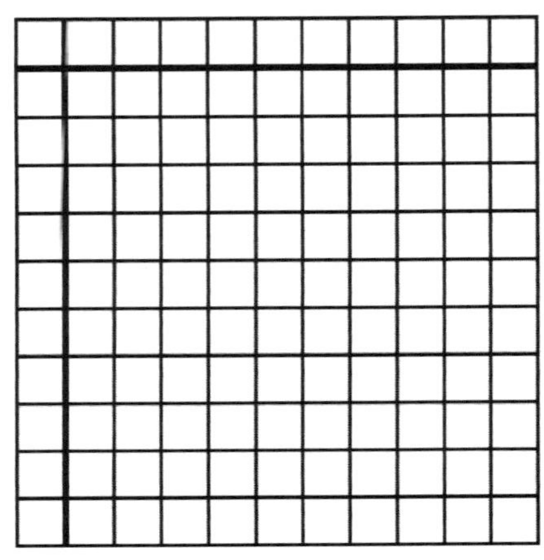

x	1	2	3	4	5	6	7	8	9	10
1	1	2	3	4	5	6	7	8	9	10
2	2	4	6	8	10	12	14	16	18	20
3	3	6	9	12	15	18	21	24	27	30
4	4	8	12	16	20	24	28	32	36	40
5	5	10	15	20	25	30	35	40	45	50
6	6	12	18	24	30	36	42	48	54	60
7	7	14	21	28	35	42	49	56	63	70
8	8	16	24	32	40	48	56	64	72	80
9	9	18	27	36	45	54	63	72	81	90
10	10	20	30	40	50	60	70	80	90	100

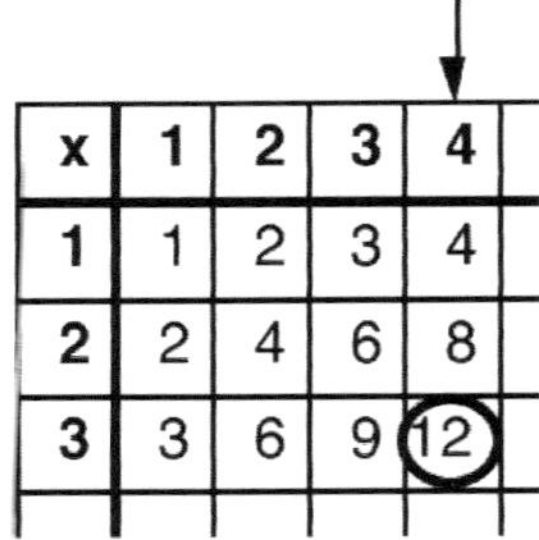

x	1	2	3	4
1	1	2	3	4
2	2	4	6	8
3	3	6	9	(12)

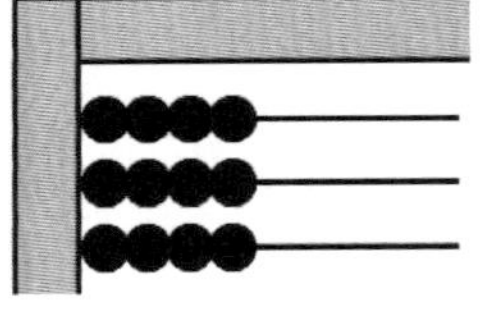

4 + 4 + 4 = _______

3 + 3 + 3 + 3 + 3 + 3 + 3 = _______

1 + 1 + 1 + 1 + 1 + 1 + 1 + 1 = _______

2 + 2 + 2 = _______

7 + 7 + 7 + 7 + 7 = _______

5 + 5 = _______

6 + 6 + 6 + 6 = _______

5 + 5 + 5 + 5 + 5 + 5 = _______

9 + 9 + 9 + 9 = _______

$$13 = 6 + 7$$

Write on the board various combinations and ask if they are true:

$$4 = 2 + 2 \text{ [yes]}$$
$$5 = 5 \times 0 \text{ [no]}$$
$$6 = 10 - 5 \text{ [no]}$$
$$7 = 7 \times 1 \text{ [yes]}$$

Next tell the children that now they are going to play a type of guessing game. Write on the board

$$10 = 9 \_ 1$$

and ask the them what they think the missing sign is. Whatever the answer, write it in and check by performing the calculation. Continue with their guesses until the correct answer is found.

Have them try other examples, such as $2 \_ 3 = 6$ [x] and $8 \_ 9 - 1$ [=] and even one with two signs missing that has two answers, $8 \_ 5 \_ 3$ [=, + or -, =]. Then give them a worksheet (8-20).

MULTIPLICATION RESULTS

The next group of topics includes the commutative and associative laws. The distributive law will be deferred until Unit 10.

Write the equation

Writing an equation to show the quantity of an array is important as a prelude to finding area. Begin by entering an array on the abacus and asking the children to state the equation. Show them 6×5 and ask them to state the equation. [$6 \times 5 = 30$] Give them other examples, such as 3×9 and 8×4.

Have the children work with partners with one entering an array and the other stating the equation. Give them a worksheet (8-21) with pictures of configurations of the abacus for which they write the equations.

Next present to them an array made with rectangles such as the 5 by 3 as shown in the figure. Ask them for the equation showing the array. [$5 \times 3 = 15$]

If available, give them pictures of arrays (8-22) for which they can write the equations, including the solutions.

ORAL PROBLEMS. A. Danielle pasted 8 rows of stamps with each row having 7 stamps. Cody pasted 7 rows of stamps. Each of his rows had 8 stamps. Who pasted more stamps? [the same]

B. Conrad set the table for 8 places. For each place he set 4 pieces of silverware. How many pieces of silverware did he set altogether? [32 pieces]

$10 = 9 \_ 1$
$8 \_ 5 = 40$
$2 = 7 \_ 5$
$1 \_ 1 = 0$
$12 \_ 10 + 2$
$0 = 8 \_ 0$
$8 \_ 2 = 10$
$10 \times 7 \_ 70$
$9 = 3 \_ 6$
$10 \_ 10 = 100$
$10 = 2 \_ 8$
$1 \_ 5 = 5$
$9 = 7 \_ 2$

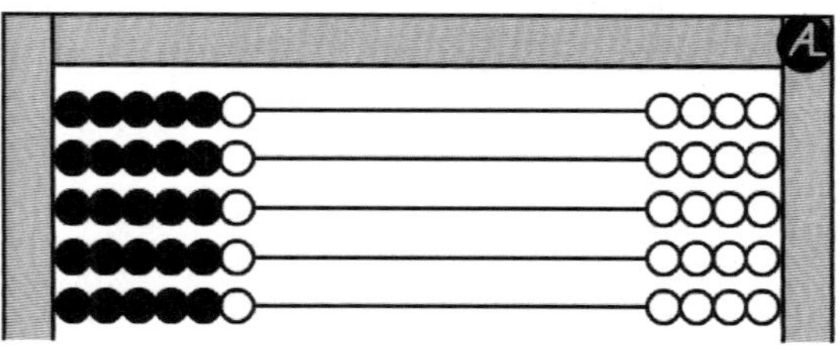
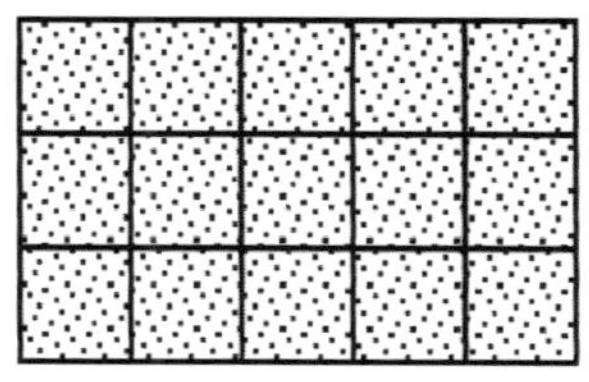
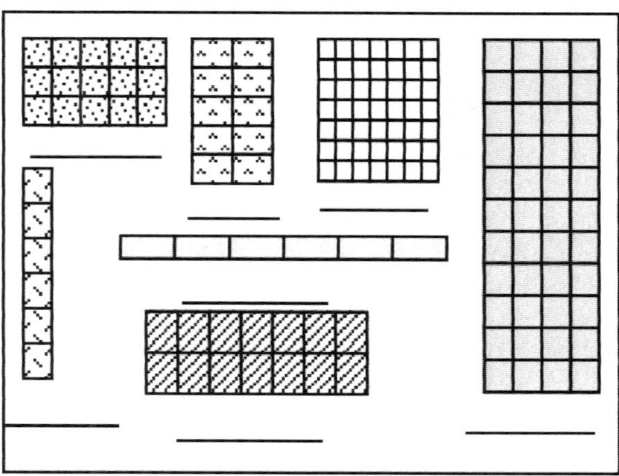

Changing the order (commutative law)

By now many of the children will have discovered the commutative law. Give them a worksheet (8-23) with pairs, as shown, before discussing the outcome. Tell them to find the rule.

After they have completed the worksheet, ask them if they know the rule. [The order does not matter.] Ask if the order matters in addition, [no] in subtraction? [yes]

To help them see the law in another manner, enter 2×6 on an abacus and ask the children to state the equation. Ask one of them to write it for all to see. $[2 \times 6 = 12]$ Next turn the abacus sideways as shown and again ask a child to state the equation. $[6 \times 2 = 12]$

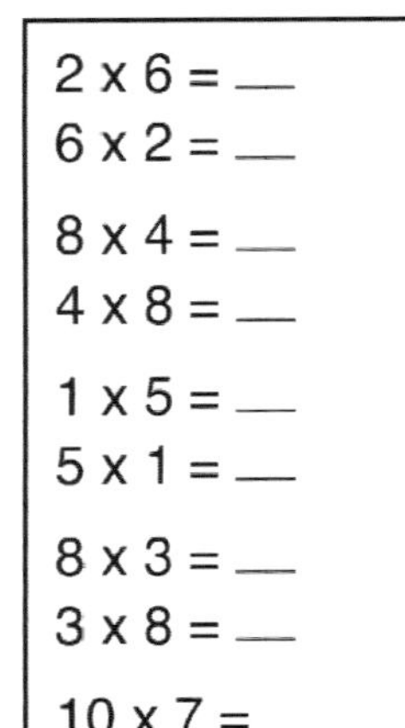

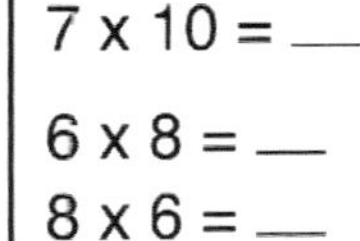

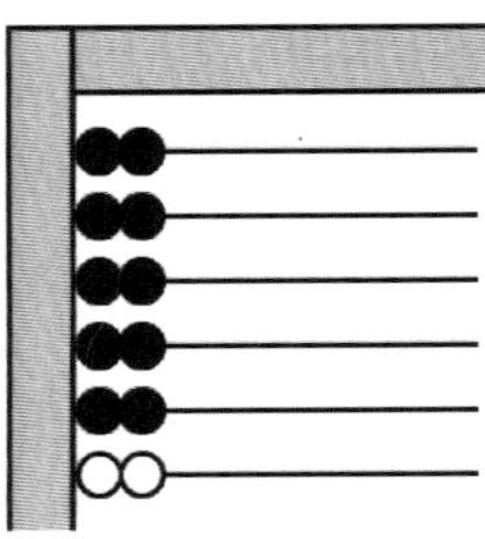

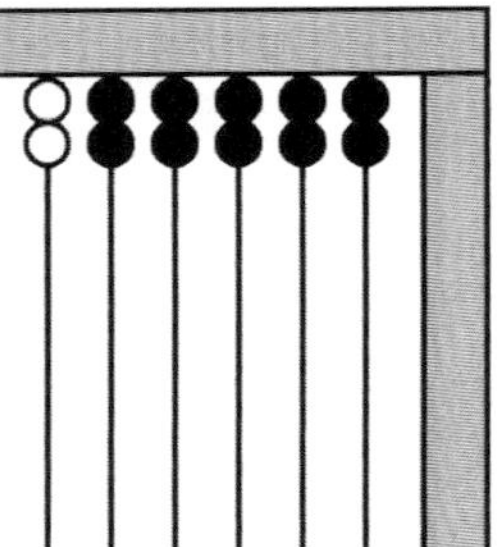

Changing the grouping (associative law)

1. Tell the children that they are going to see what happens when three numbers are multiplied together. Write on the board

$$5 \times 2 \times 3 = \underline{}$$

and explain that it can mean 5 times 2 taken 3 times. Ask a child to enter 5 times 2. Ask another child to enter another 5 times 2 on the same wires but after a small gap. Then ask a third child to enter the last 5 times 2 but to leave a gap from the edge so all three groups are separate. See the figure. The quantities can be combined as usual for determining the answer.

Ask the children how they would like to change the order of the numbers. Write down one of their suggestions and repeat the procedure ($3 \times 2 \times 5$ and $2 \times 5 \times 3$ are shown) as above. Then ask them to compare the answers. Repeat for a third time. Then give them worksheet (8-24).

THOUGHT QUESTION. How many different ways could those three numbers be rearranged? [six]

2. After the children have done the worksheet, ask if the order makes a difference. Another way to help the children understand this concept is with cubes, especially interlocking cubes. For example, build $5 \times 2 \times 3$ with three rectangles of 5 by 2 and then stack the rectangles. Other children could build the other combinations. Let the children discover that all the stacks are identical if rotated appropriately.

Once the concept is completely understand, write on the board

$$10 \times 3 \times 2 = \underline{}$$

and ask, **What is the easiest way to find the product?** Guide them to multiply the 3 and 2 first because multiplying by 10 is easy. Repeat with $5 \times 7 \times 2$. Here help them discover the 5×2 would be easiest. Give them a worksheet (8-25) with this type of problem.

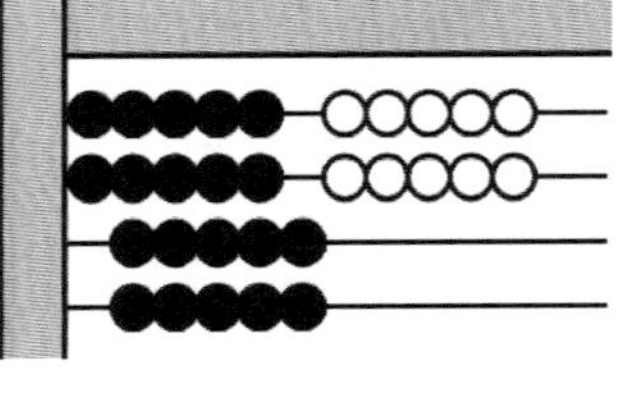

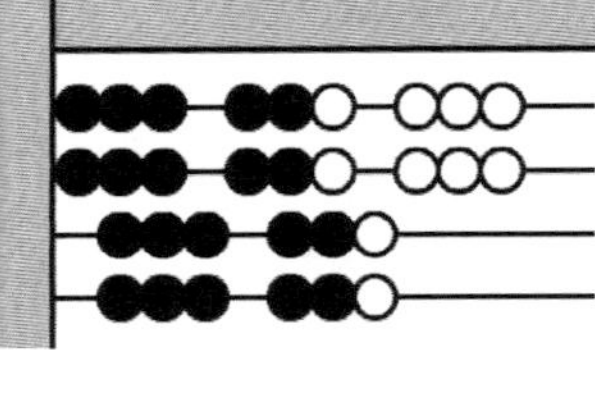

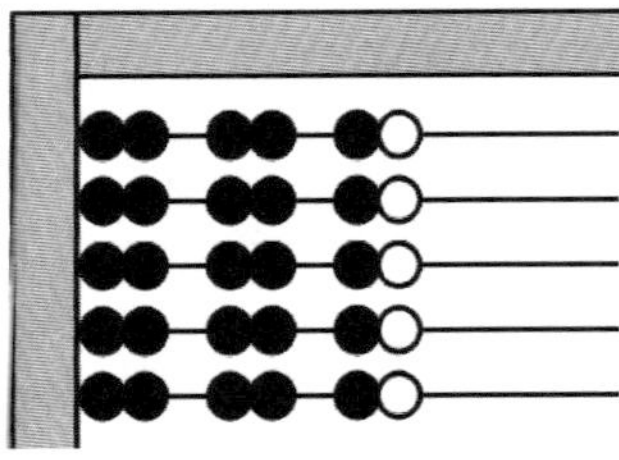

Mastering the facts

Before working with multiplying several digits, the children will need to review the facts and to learn strategies for those they do not know by memory. Since the children have had multiplication presented to them using different models, they will have visual memories of many of the facts. However, it is not possible to present work in the higher decades for further practice, as was done with addition and subtraction.

Tell them that either number being multiplied is called a FACTOR. Ask, **What are the factors in this equation, 9 × 6 = 54.** [9 and 6]

Tell them that today they will learn several tricks to help them learn their facts. Write on the board

        even × even =
        even × odd =
        odd × odd =

and ask the children to figure out the answers using examples from the multiplication table. Guide them into summarizing the results that the products are even unless both factors are odd. Ask, **Could 5 × 8 be an odd number?** [no] **Could 7 × 9 be an even number?** [no]

Give them worksheets (8-26 to 8–29) with the facts to work without the abacus in order to identify the facts they do not know. Some strategies that the children might find helpful are given below. Ask the children to look at the patterns to verify these strategies.

NINES. Nine times a number will be 1 less in the tens place than the number multiplied. For example, 9 × 4 has 3 in the tens place. Also, the two digits added together always equal 9.

EIGHTS. In the first row, the 8s products also have 1 less in the tens place than the number multiplied. For example, 8 × 3 has a 2 in the tens place. However, the second row has 2 less. Try 8 × 9 and note that the tens place of the product is 2 less than 9, or 72.

EVEN FIVES. For the even 5s, the tens place is half of the number multiplied. That is, 6 × 5 is 30, the 3 is half of 6. A odd fact can be calculated by adding 5 to the previous even fact.

EVEN SIXES. The ones of the even 6s are the same as the factor. The 4s products can be thought of a double of the product for the corresponding 2s.

FACTS OVER 5 × 5. Almost everyone will enjoy this method for determining facts greater than 5 × 5. On the bottom two wires enter the two numbers. The tens digit will be the *sum* of the dark beads at the left. The ones digit is the *product* of the number of dark beads at the right on the two wires. For example, to find 7 × 8, enter 7 and 8 on the last two wires as shown. The tens digit is 50, the number of dark beads is 2 and 3. The ones digit is 3 × 2,

1 x7	2 x5	4 x4	9 x7	4 x5
8 x1	2 x2	0 x6	2 x8	5 x6
5 x9	3 x7	6 x3	10 x2	1 x1
5 x10	3 x6	5 x4	1 x5	6 x8
7 x9	10 x10	9 x2	2 x0	3 x1

```
even x even = even
even x odd = even
odd x odd = odd
```

9	18	27	36	45
90	81	72	63	54

8	16	24	32	40
48	56	64	72	80

5	10
15	20
25	30
35	40
45	50

6	12	18	24	30
36	42	48	54	60

2	4	6	8	10
12	14	16	18	20

4	8	12	16	20
24	28	32	36	40

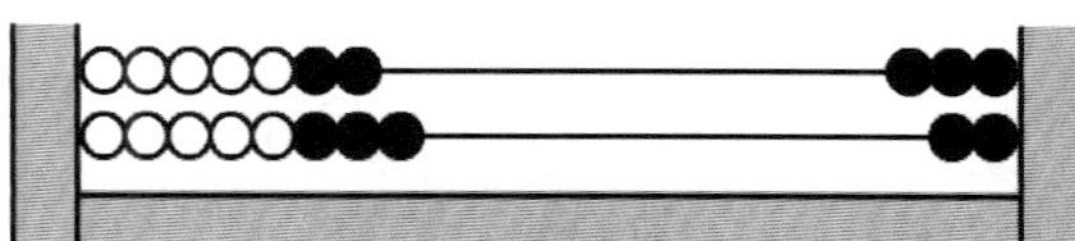

or 6, the product of the number of dark beads on each of the two wires at the right.

Actually, this scheme will work for numbers less than 5 although it is not practical. However, one additional rule is needed: subtract 10 for each light bead at the right.

FUN FACTS. And finally, two curious facts are $12 = 3 \times 4$ and $56 = 7 \times 8$.

To provide for extra practice, use the math card games.

MULTIPLICATION ALGORITHM

The best way to understand the multiplication algorithm is through successive addition. The comparison to addition will make clear why the number carried is added after the next multiplication. Most of the worksheets can be done with the children working in pairs, one calculating with the abacus and the other recording the results. Working together in pairs requires talking aloud about the work, further helping in understanding.

Multiplying tens

This lesson on multiplying multiples of tens is one in which the children can do the worksheet to discover the rule. Give them a sample problem on the board, such as

$$\begin{array}{r} 20 \\ \times 6 \\ \hline \end{array}$$

Ask them to turn their abacuses to side 2 of the abacus and ask them what 20×6 means. [20 added 6 times] Guide them to enter 20 six times, to combine, and to trade giving an answer of 120.

Give them problems (8-30) to work with the abacus and problems to work without the abacus.

ORAL PROBLEMS. A. Rick's school has 6 female gerbils. If each gerbil has a litter of 9 babies, how many babies would the school have? [54 babies]

B. A certain building has 8 floors. Each floor has 80 windows. How many windows does the building have? [640 windows]

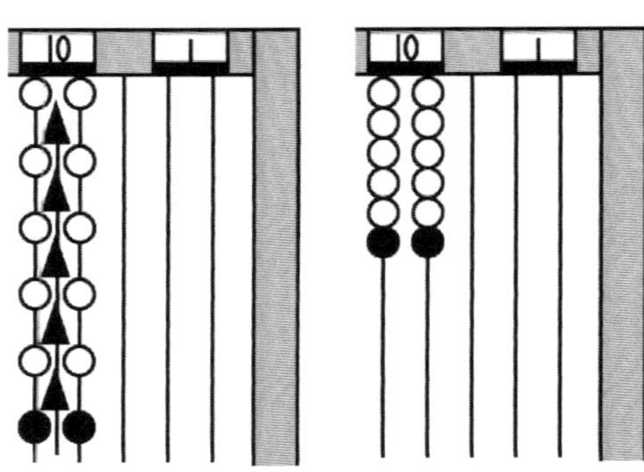

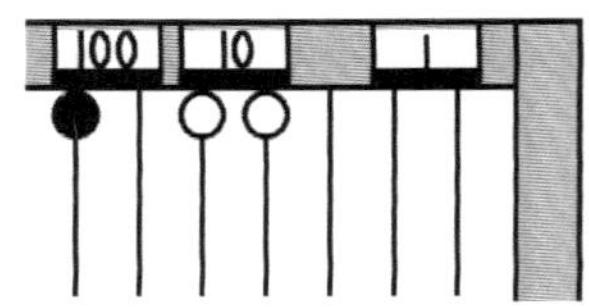

20	30	60
x6	x5	x3
40	30	50
x3	x6	x4
90	40	50
x2	x5	x3

Multiplying 2-digit numbers

1. Tell the children that they are now ready to multiply larger numbers. Write on the board

$$14 \times 5 =$$

and ask them to think about how they would multiply these numbers using the abacus horizontally. If necessary, ask the following questions, **How many 10s do you need?** [5] **How many 4s?** [5] **What is the product?** [70]

The figures show the preferred solution. (Another solution would be to add 5 fourteen times.) Notice that the 10s are multiplied first, necessary for efficient mental multiplication.

Give them 12 × 4 and 17 × 3 as additional examples before giving them a worksheet (8-31). This worksheet also can be used later as practice for mental work.

ORAL PROBLEMS. A. If a duck can fly 10 miles in an hour, how many miles can it fly in 8 hours? [80 miles]

B. There are 12 inches in a foot. How many inches are in 5 feet? [60 inches]

C. Jerahn bought 7 cards at 14 cents each. How much did he pay? [$0.98]

D. How many donuts are in 3 dozen? [36 donuts]

E. How many eggs are in 6 dozen? [72 eggs]

F. Jacqueline read 10 more pages than her friend who read 97 pages. How many pages did Jacqueline read? [107 pages]

G. Jacob had 13 miniature cars and Ben had 17 cars. How many cars would Ben have to give to Jacob so they would each have the save number? [2 cars, so they each have 15 cars] (This can be solved with a take and give procedure.)

2. While the horizontal model provides a good basis for understanding multiplication, it is necessary to return to the more abstract vertical form, which leads to the common algorithm.

Write on the board

$$\begin{array}{r} \overline{2}4 \\ \times\,5 \\ \hline \end{array}$$

and ask the children to add 24 five times with the abacus vertical. Before a trade is made, remind them to write the number traded above the next column.

Give them a worksheet (8-32A) with similar problems.

ORAL PROBLEMS. A. For a picnic, Ms Jones bought 12 six-pack cans of soda pop. How many cans of soda pop did she buy? [72 cans]

B. Beau worked 17 hours at $2 an hour. How much money did he make? [$34]

C. Kayla's family drove 541 miles on Monday and 343 miles on Tuesday. How much farther did they drive Monday? [198 miles]

D. Louis counted 14 cars parked on his block. How many tires are on the cars? [56 tires]

E. Rebecca washed 14 windows. Each window had 4 panes. How many panes did she wash? [56 panes]

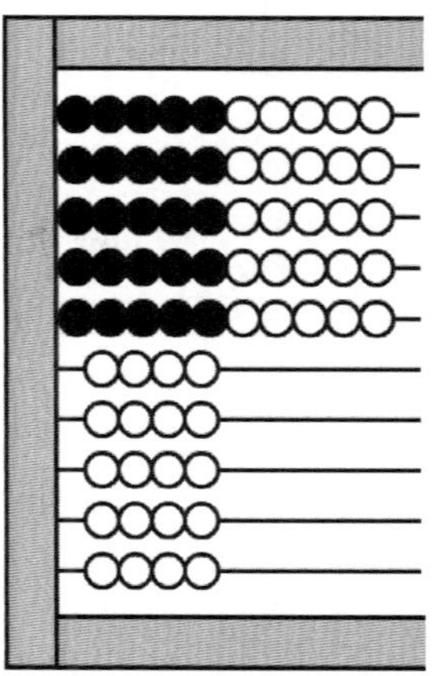
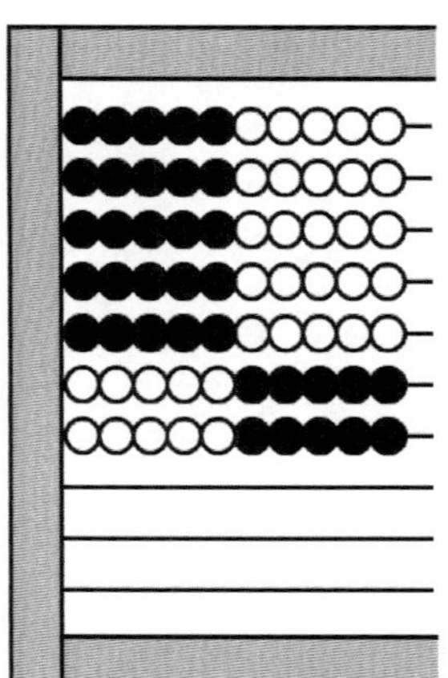

14 x 5 = __
12 x 3 = __
25 x 3 = __
36 x 2 = __
18 x 4 = __
13 x 5 = __
31 x 3 = __
19 x 4 = __
17 x 5 = __
27 x 3 = __
22 x 5 = __

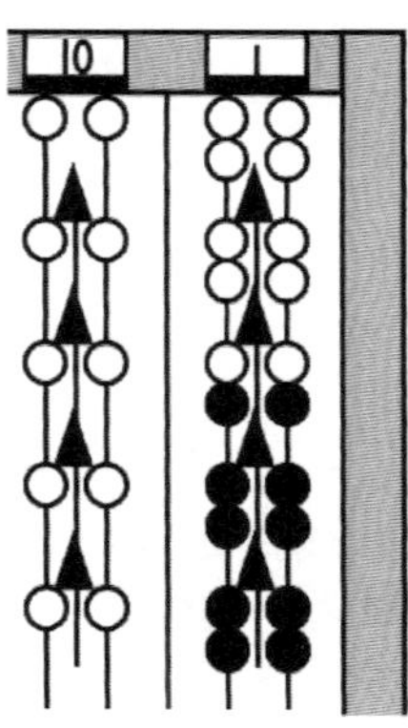
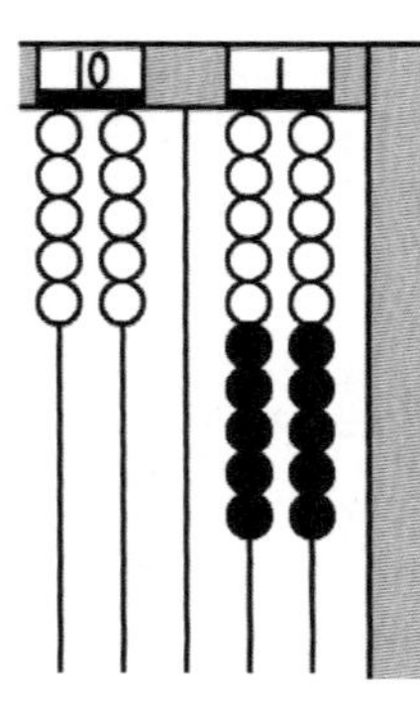

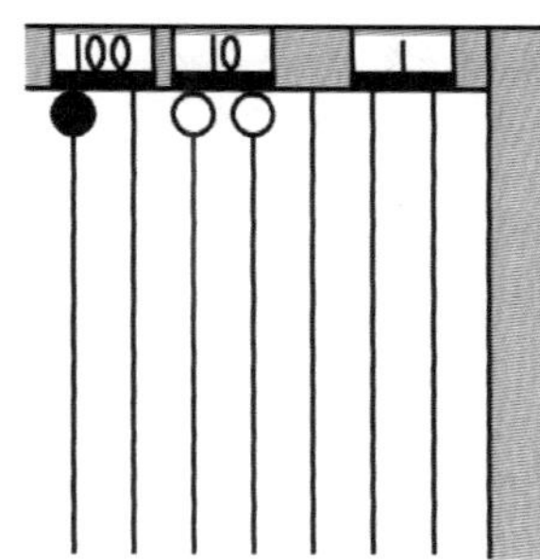

24	26	35
x 5	x 3	x 3
59	83	32
x 2	x 2	x 6
38	45	34
x 2	x 4	x 4

3. Write on the board

$$\begin{array}{r} 39 \\ \underline{\times 5} \end{array}$$

and ask the children if they are ready for a shortcut for this type of problem. Ask, **How could you enter all the 9s at one time?** [by entering 45] **What do you write above the 3?** [4] **How could you enter all the 3 tens?** [by entering 15 tens] See the figures.

Ask them to work 65 × 7 and 28 × 4 before giving them independent work (8-32B and 8-33).

Multiplying 3-digit numbers

1. Multiplying a 3-digit number on the abacus takes only one step beyond the process of multiplying a 2-digit number. Start with the problem

$$\begin{array}{r} 768 \\ \underline{\times 2} \end{array}$$

[1536] and ask the children to try doing it themselves first. The figures show the steps. Give help where needed.

Guide them through problems 126 × 5 and 318 × 7 before giving them a worksheet (8-34).

ORAL PROBLEMS. A. Every day Collin's cat eats a can of cat food that costs 51 cents. What is the cost for a week? [$3.57]

B. A school year is about 36 weeks long. How many days is that? Remember that a school week has only 5 days. [180 days]

C. How many hours are in one week? One day has 24 hours. [168 hours]

2. Some of the children may have discovered how to do these problems without the abacus. A worksheet (8-35) with columns of the same number to add may help them consolidate their understanding of multiplication as related to addition. Let them choose how to do the additions. If a child works strictly by adding, guide him or her to use the multiplication facts to make the addition easier.

ORAL PROBLEMS. A. Rachel is celebrating her eighth birthday. How many days old is she? [2920 days]

B. How long until she will be 3000 days old? [80 days]

C. Brandon just celebrated his ninth birthday. How many months old is he? [108 months]

D. Megan is 7 years old. How many weeks old is she? There are 52 weeks in a year. [364 weeks]

E. If a person eats 3 meals a day, how many meals does the person eat in a year? [1095 meals]

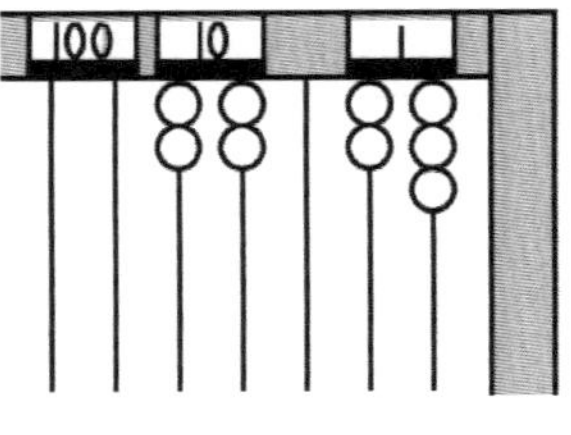

39	26	35
×5	×3	×3

59	83	32
×2	×2	×6

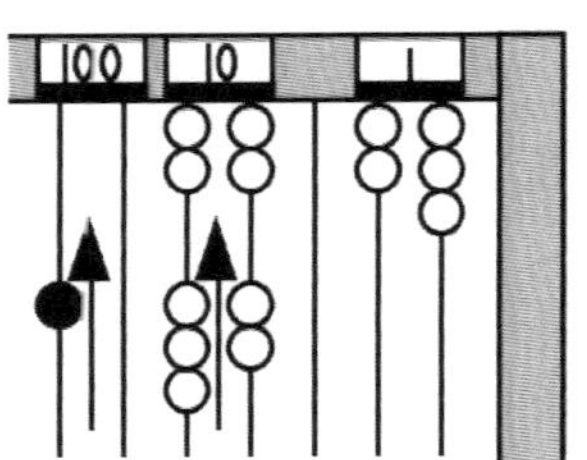

38	45	34
×2	×4	×4

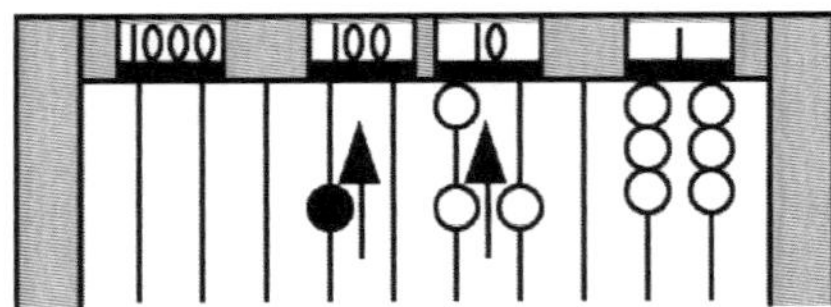

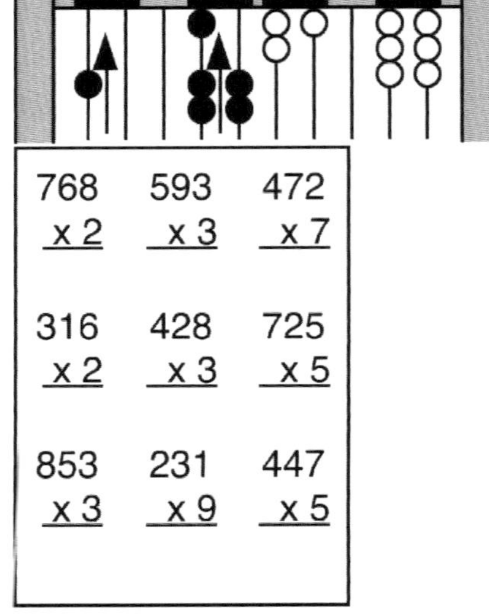

768	593	472
×2	×3	×7
316	428	725
×2	×3	×5
853	231	447
×3	×9	×5

52	17	44	89	75	36
52	17	44	89	75	36
52	17	44	89	75	36
+52	17	44	+89	75	36
	+17	44		75	+36
		+44		+75	
903			209		
903	348		209		667
903	348	870	209	508	667
903	348	870	209	508	667
903	348	870	209	508	667
903	348	870	209	508	667
+903	+348	+870	+209	+508	+667

3. Now challenge the children to work without an abacus.
If the children need help, review with a problem written
on the board

$$\begin{array}{r} 175 \\ \times 2 \end{array}$$

by asking them to multiply as usual on the abacus. Then
ask guiding questions such as, **Where did you write
10?** [The 0 under the 2 and the 1 in the next column.]
How did you get 15? [By multiplying 7 × 2 and adding the 1] **Where did you get 3 hundred?**

Give them worksheets (8-36) to do without the abacus.

ORAL PROBLEMS. A. If each month had 30 days, how
many days would there be in a year? [360 hours]

B. Kelly planted 6 rows of flowers. Each row had 2 doz-
en marigolds. How many marigolds did she plant? [144
plants]

Multiplying by multiples of 10
How to multiply and divide by multiples of 10 is essential
for advanced work in decimals.

1. The rules for multiplying by 10s is another activity that
the children can work out for themselves. Give them a
worksheet (8-37) that calls for adding the same number
ten times. For example, 46 is added to 46. That sum is
added again to 46. This is continued until 46 has been
added 10 times.

After they have completed the worksheet, write on the
board the following:

$$\begin{array}{ccccc} 32 & 85 & 12 & 256 & 638 \\ \times\,10 & \times\,10 & \times\,10 & \times\,10 & \times\,10 \end{array}$$

and ask them to state the answers. If they are unsure of
the answers they can work the problems on the abacus by
multiplying each digit in the top number by 10. For 32 ×
10, 30 times 10 is 300 and 2 times 10 is 20, giving a total
of 320. A children can also use calculators for discovering
the rule. Remind the children to write their answers carefully in the proper column.

Alternately, the problem could be turned around and
thought of as 32 tens, which could be counted out on one
or more abacuses.

Give the children worksheets (8-38) with problems for
multiplying by 10.

ORAL PROBLEMS. A. Balloons are sold in packets of
10 each. Denila bought 13 packets for a class project.
How many balloons does the class have? [130 balloons]

B. Joshua is 10 years old. How many weeks ago was he
born? There are 52 weeks in a year. [520 weeks]

$$\begin{array}{ccc} {}^{1} & {}^{1\ 1} & {}^{1\ 1} \\ 1\,7\,5 & 1\,7\,5 & 1\,7\,5 \\ \underline{\ \times 2\ } & \underline{\ \times 2\ } & \underline{\ \times 2\ } \\ 0 & 5\,0 & 3\,5\,0 \end{array}$$

175	257	497
× 2	× 6	× 8
774	802	888
× 8	× 2	× 9
285	227	98
× 7	× 2	× 7
598	615	412
× 3	× 4	× 9

46	78
+ 46	+ 78
+ 46	+ 78
+ 46	+ 78
+ 46	+ 78
+ 46	+ 78
+ 46	+ 78
+ 46	+ 78
+ 46	+ 78
+ 46	+ 78
+ 46	+ 78

16	18	20
x10	x10	x10
25	28	30
x10	x10	x10
33	35	39
x10	x10	x10
47	59	63
x10	x10	x10

2. Write on the board

$$\begin{array}{cc} 34 & 34 \\ \underline{\times\,10} & \underline{\times\,20} \end{array}$$

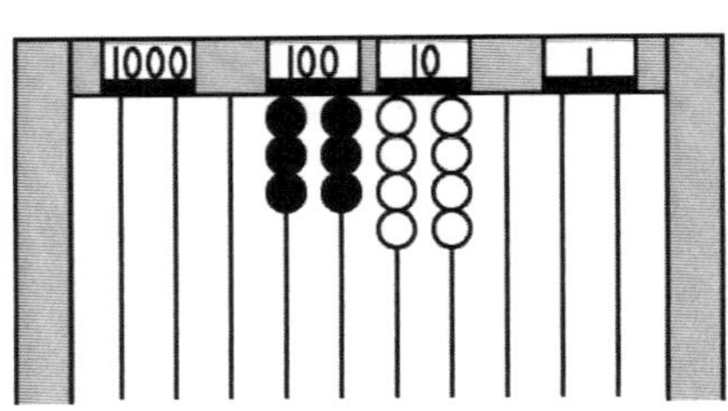

and challenge the children with, **Since you know how to find 34 times 10, how do you think you would find 34 × 20?** Guide them into thinking that it will be 2 times as much.

Show them on an abacus to start on the tens wires with 4 × 2. Next proceed with 30 × 2. See the figure. Ask a child to read the answer and ask all the children if it is twice as much as 34 × 10.

Now write on the board

$$\begin{array}{c} 74 \\ \underline{\times\,20} \end{array}$$

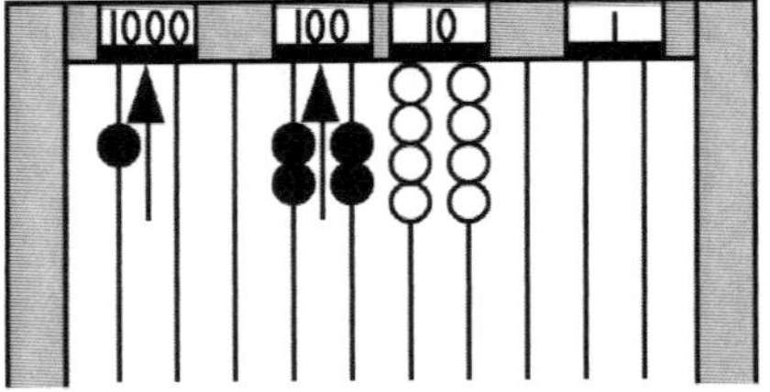

and ask the children to multiply. Remind them, **Since 74 times 2 is 10 times more than 74 × 2, we will start on the tens wires.** See the figure.

Give them 45 × 60 and 38 × 40 to multiply before giving them worksheets (8-39).

Multiplying by 2-digit numbers

1. Write on the board

$$\begin{array}{c} 34 \\ \underline{\times\,78} \end{array}$$

and tell the children that they are now ready to multiply such numbers. Ask one child to multiply 34 by 8 on the abacus and to write it down. Then ask another child to multiply 34 by 70 and write it down. A third child does the addition, trading, if necessary, and writes the product.

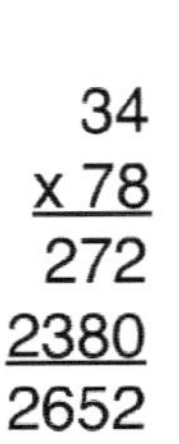

$$\begin{array}{r} 34 \\ \underline{\times\,78} \\ 272 \\ \underline{2380} \\ 2652 \end{array}$$

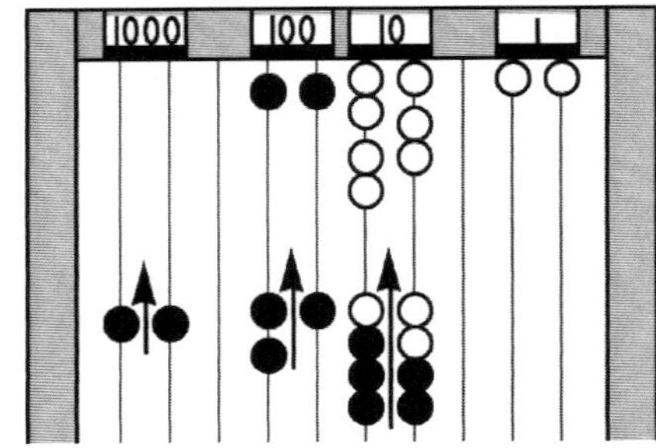

Tell the children that the products written before the answer are called PARTIAL PRODUCTS. Ask, **How many partial products does 34 × 78 have?** [2]

The 0 that occurs at the right in the second partial product is entirely optional. There is no point in teaching the children to avoid writing it; in this age of calculators, efficiency in algorithms is no longer imperative. However, a child who decides it is unnecessary may omit it.

Ask the group to work through two more examples, 29 × 56 and 56 × 29. Then give them worksheets (8-40, 8-41) to work in small groups of two or three. Most children probably will be able to do the second worksheet without the abacus. After they do the second worksheet, ask which of the two problems on a line was easier to do.

2. The children can now expand to three-digit numbers. Work problems 423 × 22, 195 × 18, and 287 × 33 with them before assigning worksheets (8-42 and 8-43).

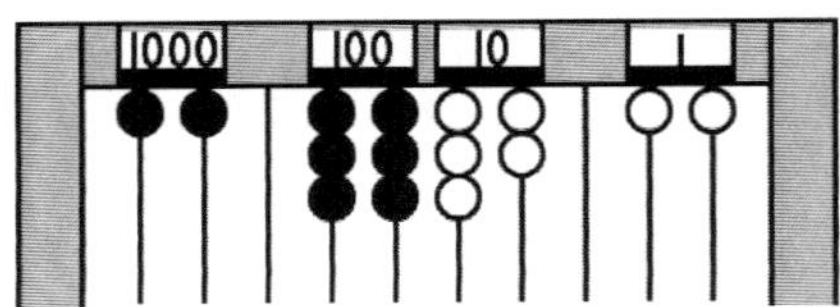

34	44	39
× 78	× 18	× 24
75	28	64
× 42	× 49	× 29

39	45
× 45	× 39
72	61
× 61	× 72
83	27
× 27	× 83
50	89
× 89	× 50

Unit 9
Division

Division answers two questions; for example, 24 divided by 3 tells either how many 3s are in 24 or how much is in each of 3 groups. In this unit, the children will be shown different ways of dividing numbers. During this work, they will memorize many of the facts. Later in the unit, the children will be tested to determine how many facts they know, and strategies can be given for those they still need to learn.

Division has a profusion of signs to represent it, namely, ÷, ⟌, —, and /. The first of these, the division sign, primarily is used to show numeric calculations in arithmetic; it is often found on calculators. The second symbol, the "doghouse," is reserved for calculating with paper and pencil. The third symbol, the dividing line, is widely used in most branches of mathematics; it is also extensively used in divisions resulting in quotients less than 1, usually referred to as fractions. The fourth symbol, the slanted line, is used in print, including recipes, on highway signs, and in work on computers.

Division by 0 is a topic that does not need to be introduced to primary students. The older child can appreciate the impossibility. Help a child to understand by asking how many 0s are in the number: it makes no sense, so we say it is impossible.

Short division is a skill that is frequently needed and should be mastered first. Long division, however, is seldom done by hand today because of the widespread use of calculators. In this unit, it is approached as an extension of short division.

INTRODUCING DIVISION

Division will be studied from several different perspectives, including counting, skip counting, and the inverse of multiplication.

Division by counting

Tell the children that now they will start division. Pose this problem, **If three friends have 12 pens that they want to share evenly, how can they share them?** Guide them to the solution of passing them out, 3 at a time, until they are gone. Repeat with other objects, such as 18 books shared evenly among 6 children.

Show 12 ÷ 3 on the abacus. Use golf tees or markers from a game to represent the persons (3 for this example). Enter beads evenly until the total quantity (12) is reached. Point to the first wire, or marker and ask, **How many did this one receive?** [4] Repeat for the second and third wires. See the figures.

Show them how to write it saying, **This is one way we write it**. Write on the board

$$12 \div 3 = 4$$

while saying, **Twelve divided by 3 equals 4.**

Guide them through two other examples, such as, 14 ÷ 2 and 20 ÷ 4 before giving them worksheets (9-1).

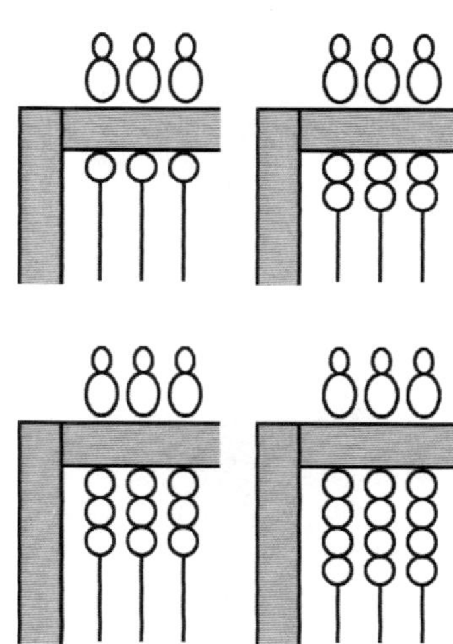

$$12 \div 3 = \underline{\quad}$$
$$10 \div 2 = \underline{\quad}$$
$$12 \div 4 = \underline{\quad}$$
$$15 \div 5 = \underline{\quad}$$
$$16 \div 4 = \underline{\quad}$$
$$21 \div 7 = \underline{\quad}$$
$$14 \div 2 = \underline{\quad}$$
$$30 \div 10 = \underline{\quad}$$
$$35 \div 5 = \underline{\quad}$$
$$8 \div 4 = \underline{\quad}$$

FUTURE REVIEW. After they have completed the worksheets, write and ask them how they would do 0 ÷ 3, **How much would each person receive?** [0]

Division by skip counting

Tell the children that there is another way to write division. Write on the board

$$4\overline{)24}$$

and explain that it is read as 24 divided by 4 and means how many 4s in 24. One way to find the answer is to enter and count by 4s until 24 is reached. The number of wires tells how many 4s are in 24. Show them where to write the answer, above the ones place of the 24.

Repeat for $3\overline{)18}$ and $5\overline{)20}$. Then give them worksheets (9-2) for practice.

Division by take and give

Division by counting and by skip counting are somewhat tedious, but many children enjoy division by take and give. It consists of rearranging a quantity into the desired array in one of two methods.

1. Write on the board

$$7\overline{)28}$$

and ask the children to enter 28. Tell them, **We want to find out how many 7s are in 28, so we want only 7 on each wire.** Ask the children to remove the extra beads by take and give. See the figures. **So how many 7s are in 28?** [4]

Repeat for $5\overline{)20}$ and $9\overline{)18}$. Then assign them similar worksheets (9-3).

2. The second method involves arranging the beads evenly on 7 wires. Say, **We want to find out how much is on each of 7 rows.** Again write on the board

$$7\overline{)28}$$

and ask the children to enter 28. Tell them that we want to distribute the 28 over 7 wires. Use take and give; start at the 7th wire with the transferring. Although 4 beads were transferred at a time in the figures below, any quantity can be transferred at a time. See the figures. The worksheet (9-4) is similar to the last one.

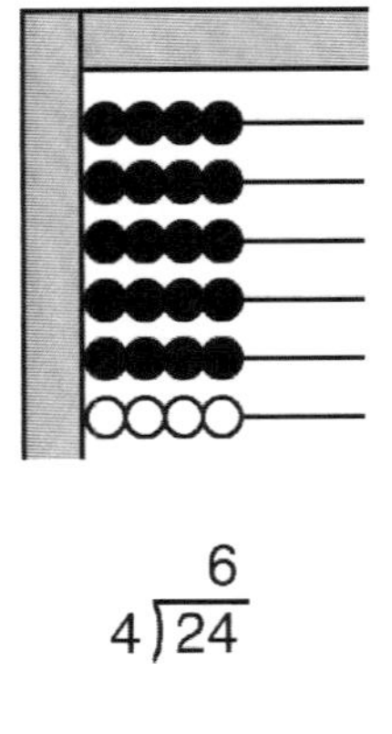

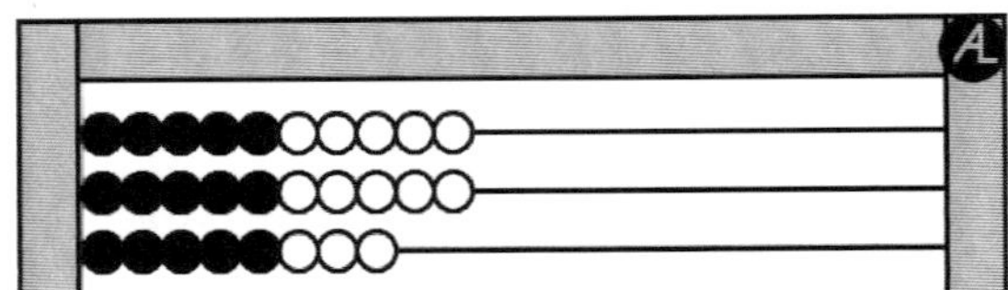

$4\overline{)24}$	$4\overline{)32}$
$5\overline{)50}$	$2\overline{)16}$
$3\overline{)12}$	$7\overline{)28}$
$4\overline{)28}$	$10\overline{)90}$
$5\overline{)20}$	$6\overline{)18}$
$8\overline{)40}$	$7\overline{)35}$
$5\overline{)15}$	$4\overline{)8}$

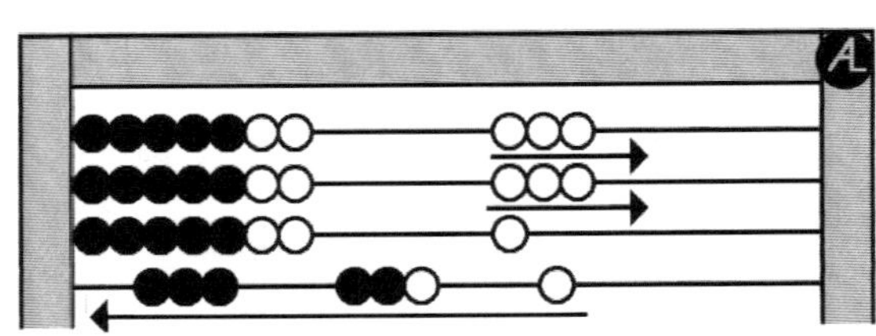

$7\overline{)28}$	$8\overline{)56}$
$4\overline{)16}$	$3\overline{)18}$
$3\overline{)21}$	$9\overline{)72}$
$6\overline{)30}$	$2\overline{)16}$
$5\overline{)25}$	$3\overline{)24}$
$8\overline{)32}$	$7\overline{)42}$
$4\overline{)28}$	$2\overline{)14}$

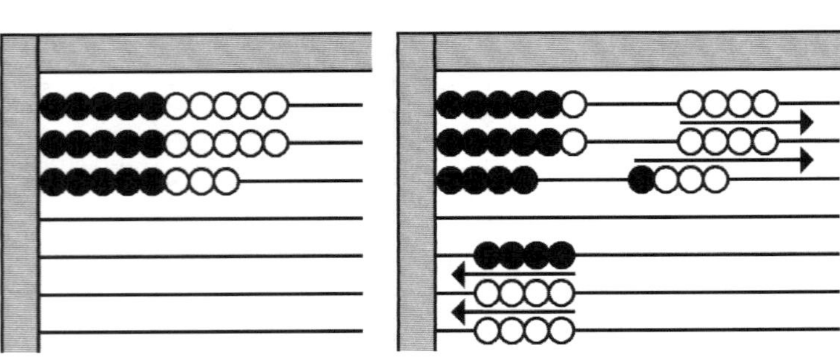

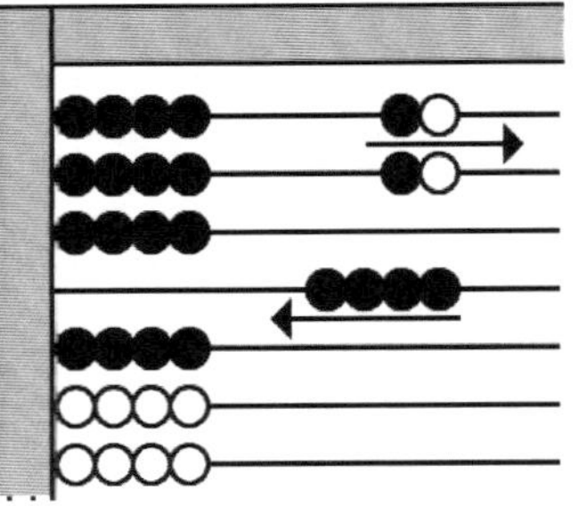

The dividing line

1. Dividing by using the multiplication table is another activity designed to help the children see the correlation between multiplication and division.

Write on the board

$$\frac{24}{8}$$

and tell the children that this is another way to write 24 divided by 8. Show them how to use the multiplication chart (see the appendix) to find the answers. Find the column that starts with 8 and look down until 24 is found. Then either count down the number of rows or look at the left to see in which row [3] the 24 is found.

Tell them that the answer in division is called the QUOTIENT. Ask, **What is the quotient in 24 divided by 8?** [3]

Give them a few other examples, such as $\frac{18}{2}$ and $\frac{42}{6}$. Assign them a worksheet (9-5).

2. This activity will teach the term dividend and will review for the children the different ways to represent division. Write on the board

$$\frac{48}{6}$$

and tell them, **The number we start with** (point to the 48) **is the DIVIDEND**. Ask a child to read the example [48 divided by 6] and to write it another way and to show the dividend. Repeat with a second child. Ask another child to write the quotients.

Ask the children to write 14 divided by 7 in three ways. Let them choose any method to find the answer. Ask, **What is the dividend** [14] **and what is the quotient?** [2]

Give them the worksheet (9-6). You might want to ask them to identify the quotients and dividends with special colors or symbols.

Repeated subtraction

Repeated subtraction really is removing groups; it demonstrates how division is related to subtraction in the same way that multiplication is related to addition.

Write on the board

$$6\overline{)24}$$

and remind them that this problem can be thought of as how many 6s are in 24. Enter 24 and explain, **We will remove 6 at a time. To keep track we will enter a 1 on the bottom wire each time we remove a 6.** Invite a different child to remove each 6 and enter the counter. The 6s can be removed in any configuration. See the figure.

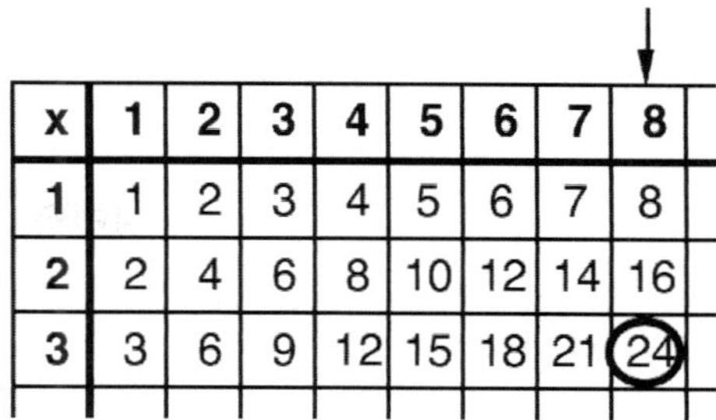

x	1	2	3	4	5	6	7	8
1	1	2	3	4	5	6	7	8
2	2	4	6	8	10	12	14	16
3	3	6	9	12	15	18	21	(24)

$$\frac{24}{8} = \underline{\quad} \qquad \frac{54}{6} = \underline{\quad}$$

$$\frac{42}{7} = \underline{\quad} \qquad \frac{24}{4} = \underline{\quad}$$

$$\frac{10}{2} = \underline{\quad} \qquad \frac{24}{6} = \underline{\quad}$$

$$\frac{16}{2} = \underline{\quad} \qquad \frac{32}{8} = \underline{\quad}$$

$$\frac{48}{6} = 8 \qquad 48 \div 6 = 8 \qquad 6\overline{)48}\,^{8}$$

		$\frac{48}{6} = \underline{\quad}$
$10 \div 5 = \underline{\quad}$		
	$9\overline{)72}$	
		$\frac{21}{7} = \underline{\quad}$
	$2\overline{)0}$	

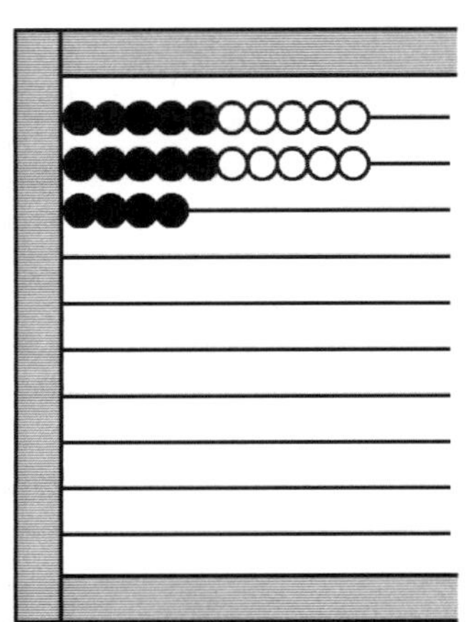 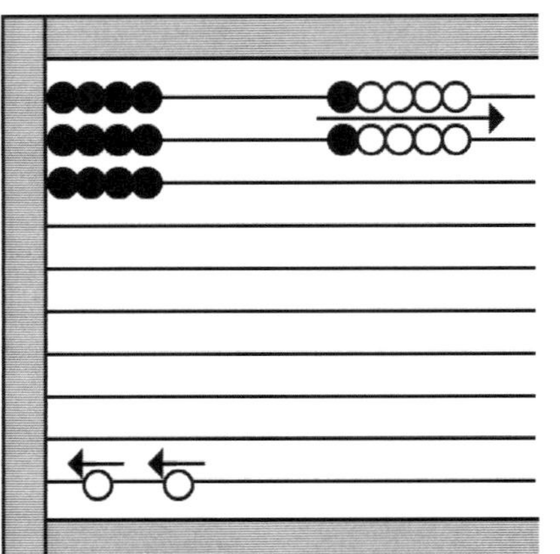 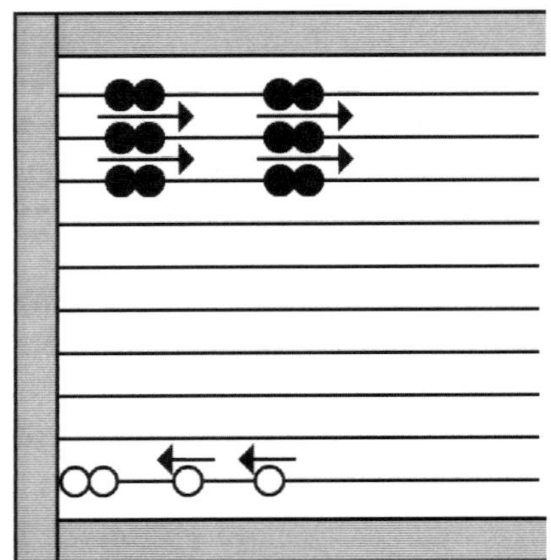

Ask the children to work in pairs with other examples, $5\overline{)25}$ and $4\overline{)32}$ and then a worksheet (9-7).

ORAL PROBLEMS. As part of problem solving, be sure the children show the problem on the abacus and write the equation. The problems given here are examples.

A. Brittany has 12 sheets of construction paper for her group. If her group has 4 members, how many sheets does each member receive? [3 sheets]

B. Alex planted 30 tulips in 6 rows. How many tulips did he plant in each row? [5 tulips]

C. Tina has 42 marigolds. She wants 6 in each row. How many rows can she plant? [7 rows]

D. Brent had 60 cents. How many balloons can he buy if each balloon costs 6 cents? [10 balloons]

E. Natasha bought 8 cards costing 72 cents. What did each card cost? [9 cents]

F. Mason counted 50 fingers at the breakfast table one morning. How many persons were there? [5 persons]

$6\overline{)24}$	$5\overline{)45}$	$9\overline{)54}$
$2\overline{)6}$	$4\overline{)16}$	$5\overline{)20}$
$6\overline{)42}$	$3\overline{)27}$	$7\overline{)56}$
$8\overline{)32}$	$7\overline{)21}$	$5\overline{)30}$
$4\overline{)12}$	$8\overline{)48}$	$7\overline{)7}$
$2\overline{)16}$	$10\overline{)90}$	$6\overline{)36}$
$5\overline{)45}$	$3\overline{)18}$	$5\overline{)35}$

DIVISION RESULTS

Following the introduction to division, the children will learn about the properties, or results, of division.

Writing equations

Writing all the possible equations for a given array will help the children see the relationship between multiplication and division: they are inverses of each other. On an abacus enter 4 rows of 9. Ask them for the two multiplication equations and the two division equations. The part/whole circles also help children see the connections. Provide help where needed.

Give them two more examples, an array of 7 rows of 5 and an array of 3 rows of 3. The second example has only two equations, 3 x 3 = 9 and 9 ÷ 3 = 3. Then give them worksheets with pictures of arrays (9-8 and 9-9).

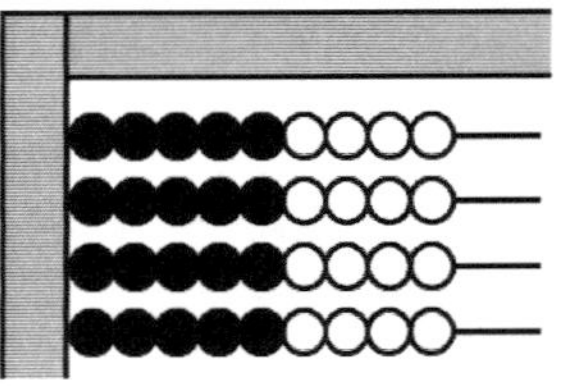

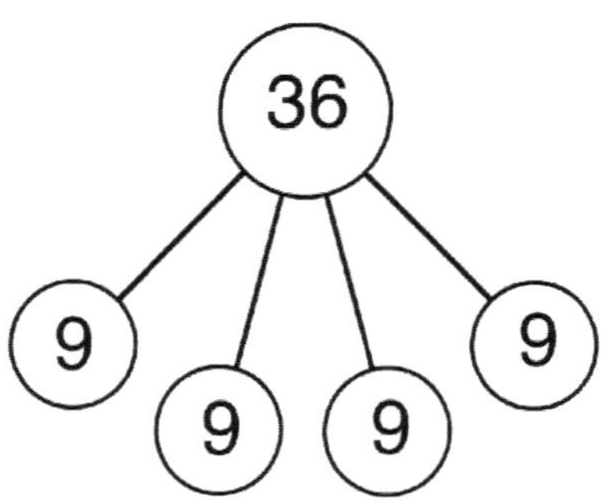

Missing factors

These missing factor equations will now make more sense to the children since they have studied division and the related equations. Write on the board

$$9 \times \_\_ = 27$$

and ask how they would solve it. After one child gives an answer, ask, **Who can think of another way to solve it?** Continue asking until all ideas are gathered.

Write

$$7 \times \_\_ = 28$$

$$\_\_ \times 5 = 40$$

and let the children use their own methods. Give them a worksheet (9-10).

$9 \times \_\_ = 27$

$4 \times \_\_ = 28$

$\_\_ \times 3 = 15$

$8 \times \_\_ = 72$

$10 \times \_\_ = 30$

$9 \times \_\_ = 54$

$5 \times \_\_ = 35$

$\_\_ \times 4 = 8$

$1 \times \_\_ = 10$

$8 \times \_\_ = 8$

Remainders

Remainders introduced too soon can interfere with the children learning the facts. However, they are an important aspect of division. As in life, things do not always turn out even.

1. Tell the children to watch for something special to happen in this division. Choose 14 identical objects, such as books, and a group of four children. Divide the books evenly among the children. After they are divided, tell them, **The extra two books cannot be divided, so they shall remain with me. They are called the REMAINDER.**

Write on the board

$$5 \overline{)32}$$

and ask the children to divide it any way they wish. Ask, **Is there a remainder?** [yes, 2] Show them one way to write it, with an "r" followed by the remainder.

Repeat for examples $10\overline{)67}$ and $6\overline{)51}$ before assigning a worksheet (9-11).

2. This next activity is also a preparation for short division. Write on the board

$$4 \overline{)28}^{7}$$

and ask the children, **What do you get when you multiply the two numbers on the outside of the "doghouse."** [28, the same number as inside] Ask, **Is this always true?** Let them come up with examples until they are convinced it is always true.

Then tell them that we can use this to find remainders. Write on the board

$$4 \overline{)31}^{7} \; r \_\_$$

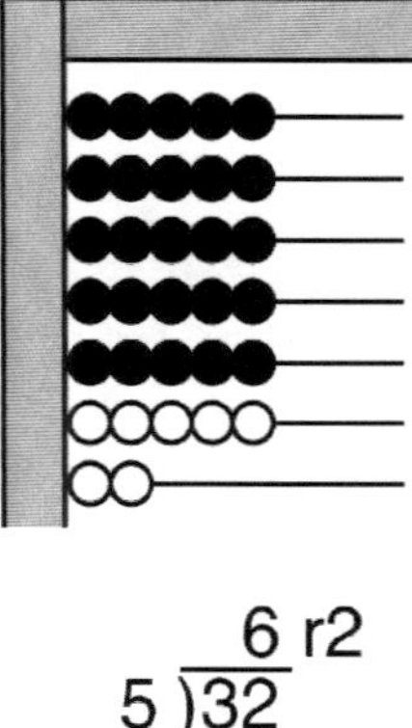

$$5 \overline{)32}^{\;6\,r2}$$

$5\overline{)32}$	$3\overline{)14}$
$8\overline{)75}$	$2\overline{)17}$
$6\overline{)41}$	$5\overline{)48}$
$6\overline{)27}$	$10\overline{)82}$
$9\overline{)49}$	$9\overline{)17}$
$6\overline{)43}$	$9\overline{)52}$
$4\overline{)41}$	$9\overline{)83}$

$4\overline{)31}^{\;7r}$	$4\overline{)14}^{\;3r}$
$9\overline{)75}^{\;8r}$	$9\overline{)56}^{\;6r}$
$6\overline{)35}^{\;5r}$	$5\overline{)8}^{\;1r}$
$3\overline{)29}^{\;9r}$	$6\overline{)50}^{\;8r}$
$2\overline{)15}^{\;7r}$	$3\overline{)32}^{\;10r}$
$5\overline{)12}^{\;2r}$	$10\overline{)37}^{\;3r}$
$4\overline{)11}^{\;2r}$	$5\overline{)38}^{\;7r}$

and ask, **How far is 28 from 31?** [3] (This subtraction is done mentally, using the counting up procedure.) **So what is the remainder?** [3] Use worksheet 9-12.

THOUGHT QUESTION. Can the remainder be more than the number it is being divided by? [no]

Summarize by saying that to find a remainder, multiply and subtract.

Increasing dividends

This activity is designed to help the children notice what happens when the dividend increases but the divisor stays the same. Give the children the worksheet (9-13), asking to work each entire row before going on.

After the worksheet, ask them what they noticed. If necessary, ask, **What did you notice about the dividends?** [They became larger.] **What did you notice about the quotients?** [They became larger.] **What did you notice about the number they were divided by?** [They stayed the same.]

Increasing divisors

This and the previous activity could have been entitled increasing numerators and increasing denominators. The work is exactly the same whether they are thought of as dividends and divisors or numerators and denominators.

Before assigning the worksheet, teach the children the term divisor. Write on the board

$$\frac{48}{6} = 8 \qquad 48 \div 6 = 8 \qquad 6\overline{)48}$$

and ask them, **What do we call the 8?** [quotient] **What do we call the 48?** [dividend] The number that 48 is divided by is called the DIVISOR. Write several examples on the board in the three forms and ask them to name the dividends [48] and the quotients [8]. Then tell them, **The number that we are dividing by, the 6, is the DIVISOR.** Ask them to name the divisors in other examples.

Give them the worksheet (9-14) as shown. After they complete it, ask them what they noticed. Guide them to observe that if the dividends stay the same, the quotients decrease as the divisors increase.

Comparing quotients

This activity reviews the < and > symbols and offers the children an opportunity to apply their observations of the previous two activities.

Review the symbols if necessary and give them the worksheet (9-15).

$$\frac{6}{2} = \underline{\quad} \qquad \frac{10}{2} = \underline{\quad} \qquad \frac{12}{2} = \underline{\quad}$$

$$\frac{9}{3} = \underline{\quad} \qquad \frac{12}{3} = \underline{\quad} \qquad \frac{21}{3} = \underline{\quad}$$

$$\frac{10}{5} = \underline{\quad} \qquad \frac{25}{5} = \underline{\quad} \qquad \frac{50}{5} = \underline{\quad}$$

$$\frac{10}{10} = \underline{\quad} \qquad \frac{60}{10} = \underline{\quad} \qquad \frac{100}{10} = \underline{\quad}$$

$$\frac{16}{8} = \underline{\quad} \qquad \frac{56}{8} = \underline{\quad} \qquad \frac{64}{8} = \underline{\quad}$$

$$\frac{54}{9} = \underline{\quad} \qquad \frac{63}{9} = \underline{\quad} \qquad \frac{81}{9} = \underline{\quad}$$

$$\frac{8}{2} = \underline{\quad} \qquad \frac{8}{4} = \underline{\quad} \qquad \frac{8}{8} = \underline{\quad}$$

$$\frac{12}{2} = \underline{\quad} \qquad \frac{12}{3} = \underline{\quad} \qquad \frac{12}{6} = \underline{\quad}$$

$$\frac{16}{2} = \underline{\quad} \qquad \frac{16}{4} = \underline{\quad} \qquad \frac{16}{8} = \underline{\quad}$$

$$\frac{24}{4} = \underline{\quad} \qquad \frac{24}{6} = \underline{\quad} \qquad \frac{24}{8} = \underline{\quad}$$

$$\frac{36}{4} = \underline{\quad} \qquad \frac{36}{6} = \underline{\quad} \qquad \frac{36}{9} = \underline{\quad}$$

$$\frac{30}{5} = \underline{\quad} \qquad \frac{30}{6} = \underline{\quad} \qquad \frac{30}{10} = \underline{\quad}$$

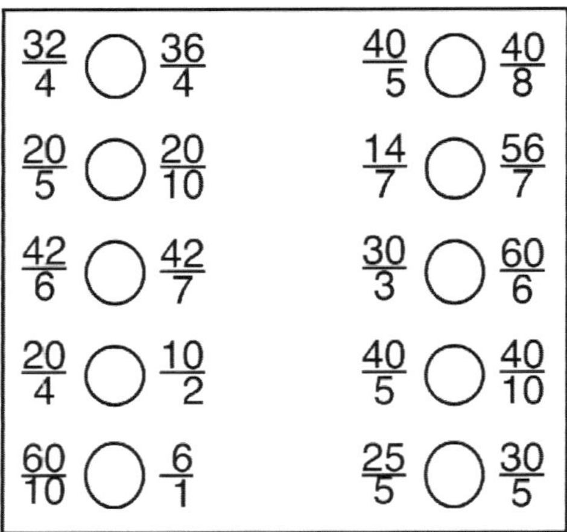

$$\frac{32}{4} \bigcirc \frac{36}{4} \qquad \frac{40}{5} \bigcirc \frac{40}{8}$$

$$\frac{20}{5} \bigcirc \frac{20}{10} \qquad \frac{14}{7} \bigcirc \frac{56}{7}$$

$$\frac{42}{6} \bigcirc \frac{42}{7} \qquad \frac{30}{3} \bigcirc \frac{60}{6}$$

$$\frac{20}{4} \bigcirc \frac{10}{2} \qquad \frac{40}{5} \bigcirc \frac{40}{10}$$

$$\frac{60}{10} \bigcirc \frac{6}{1} \qquad \frac{25}{5} \bigcirc \frac{30}{5}$$

Mastering the facts

Quick recognition of the division facts is not sufficient for dividing larger numbers. In dividing by 8, for example, one must recognize that 64, as well as 65, 66, 67, 68, 69, 70, and 71 will give 8 as the quotient, although all but 64 have a remainder.

The best strategy for division facts is the guess and check method. Ask the children to guess the answer and then check it by multiplying. The remainder is found by counting up to the dividend. It should be checked to be sure it is less than the divisor.

Give the children the division facts (9-16 to 9-18) to check which facts need additional work.

Here are some specific strategies for exact quotients:

TWO. Think of half of the number.

FOUR. To divide by 4, take half of the number twice. Try, for example, 28/4; half of 28 is 14 and half of that is 7, the answer.

FIVE. For a number ending in 0, double the tens digit when dividing by 5. Thus, 30/5 is 3 doubled, or 6. If the number ends in 5, double as before but add 1. For example, 45/5 is 4 doubled plus 1, or 9.

SIX. For dividing by 6, the quotient is either the same as the ones digit or 5 needs to be added or subtracted. For example, 48/6 is 8, the same as the ones digit. But 42/6 cannot be 2, so it is 2 + 5, or 7. Likewise, 18/6 cannot be 8, so it is 8 − 5, or 3.

EIGHT. To find the quotient for a number 40 or less being divided by 8, add 1 to the tens digit. If the number is over 40, add 2. Thus, 24/8 is 1 more than 2, or 3, and 72/8 is 2 more than 7, or 9.

NINE. Add 1 to the tens digit to find the quotient when the divisor is 9. Therefore, 63/9 is 6 plus 1, or 7.

NUMBER ENDING IN FIVE. For numbers ending in 5 where the divisor is not 5, the quotient is always 5. However, 25/5 is also 5.

DIGITS TOTALING 9. A number whose digits total 9 will have either the divisor or the quotient equal to 9. One number has both, 81/9.

Dividing larger numbers

Inform the children that now they are ready to divide larger numbers. Write on the board

$$3\overline{)42}$$

and enter 42. First let them try to discover a way to find the solution. If necessary, point out the group of 3 tens and ask, **How many groups of 3 are in 30?** [10] Remove the 30 and enter 10 on the bottom wire. Ask,

63 ÷ 9 = __	7 ÷ 7 = __	9 ÷ 9 = __
25 ÷ 5 = __	81 ÷ 9 = __	24 ÷ 8 = __
28 ÷ 7 = __	24 ÷ 6 = __	24 ÷ 3 = __
12 ÷ 4 = __	30 ÷ 6 = __	56 ÷ 7 = __
15 ÷ 5 = __	1 ÷ 1 = __	54 ÷ 9 = __
20 ÷ 5 = __	6 ÷ 3 = __	16 ÷ 2 = __
21 ÷ 3 = __	48 ÷ 6 = __	2 ÷ 1 = __
32 ÷ 8 = __	30 ÷ 5 = __	45 ÷ 9 = __
27 ÷ 9 = __	14 ÷ 7 = __	42 ÷ 6 = __

x	1	2	3	4	5	6	7	8	9	10
1	1	2	3	4	5	6	7	8	9	10
2	2	4	6	8	10	12	14	16	18	20
3	3	6	9	12	15	18	21	24	27	30
4	4	8	12	16	20	24	28	32	36	40
5	5	10	15	20	25	30	35	40	45	50
6	6	12	18	24	30	36	42	48	54	60
7	7	14	21	28	35	42	49	56	63	70
8	8	16	24	32	40	48	56	64	72	80
9	9	18	27	36	45	54	63	72	81	90
10	10	20	30	40	50	60	70	80	90	100

7	14	21
28	35	42
49	56	63
70		

8	16	24	32	40
48	56	64	72	80

9	18	27	36	45
90	81	72	63	54

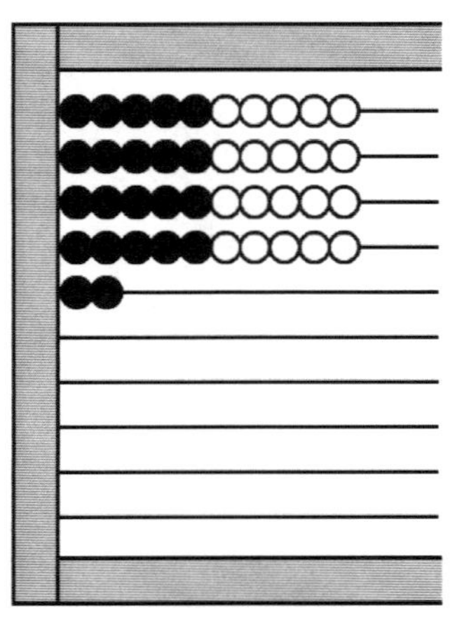 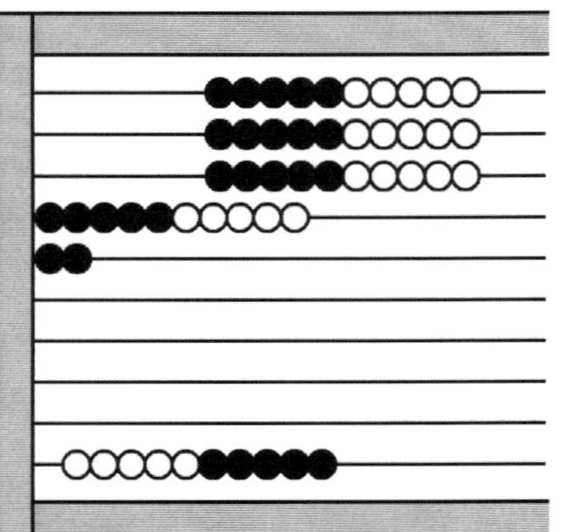 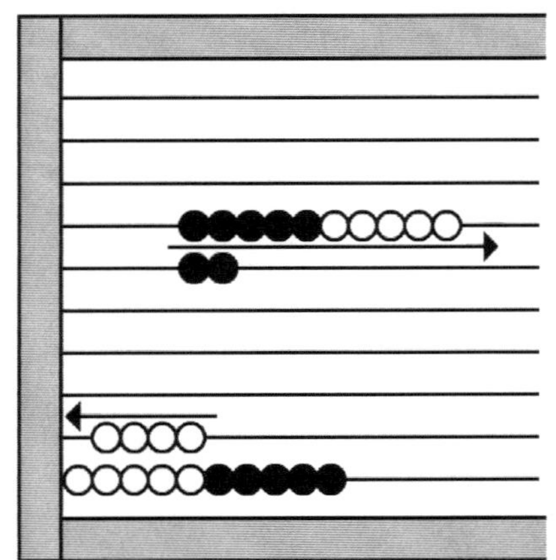

What is left? [12] **How many groups of 3 are in 12?** [4] **So what is 42 divided by 3?** [14] Instruct them when they write the answer to line it up with the dividend; that is, the 1 is written directly above the 4 and the 4, directly above the 2.

Repeat the process for 5)60 and 4)66, which has a remainder. Give them a worksheet (9-19).

SHORT DIVISION

Short division is much more important than long division and is easily mastered first. The abacus will be used in the vertical position in preparation for learning the short division algorithm.

3)42	4)72	5)70
2)36	5)80	6)72
4)56	3)54	3)60
2)30	4)60	4)64
5)75	3)66	2)50
2)53	5)68	4)62

Division without trading

For children who do not know all the division facts, especially those involving remainders, let them use a multiplication chart. The quotient is found by looking in the appropriate column for the largest multiple that is not more than the number.

1. For division on the abacus, push the beads away from the bottom about 1 ½ inches (4 cm) to allow space to enter the quotient. Write on the board

$$3)\overline{69}$$

and enter 69 as shown. Explain that the beads along the bottom must be pushed up to make room for the quotient.

Tell the children that we start at the left, the same as subtraction. First ask, **What is six tens divided by 3?** [2 tens] Remove the 6 and enter the 2 at the bottom. Ask a child to write the 2 above the 6. Then, **Nine ones divided by 3 is?** [3] Ask another child to write the 3 above the 9. **Now let's check our answer. Is 23 times 3 equal to 69?** [yes]

For the second example, let the children try 2)864. For the third example, give them 3)969. Then give them the worksheet (9-20).

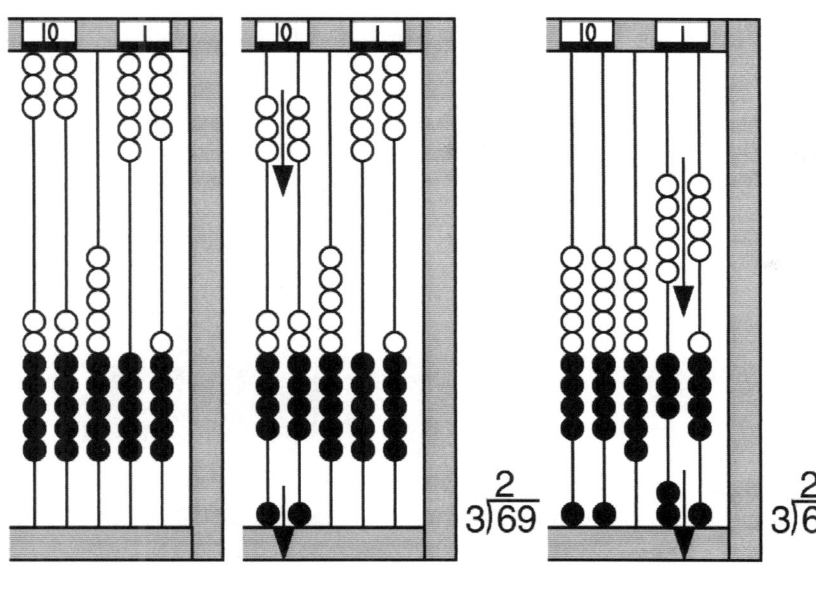

3)69	2)82	3)96
4)84	2)46	2)68
4)448	3)639	2)486
5)555	4)884	3)639
2)426	3)393	2)842
4)488	3)669	2)680

2. The next work will be on problems where the first digit is less than the divisor. Write on the board

4)128

and enter 128 on the abacus. Ask, **Can we divide the hundreds by 4?** [no] **Let's think about trading the 1 hundred for tens. How many tens would we have?** [12] Do not actually perform this trade. Rather place your fingers on the hundreds and tens wires to help them see the 12, **We do not have to trade; we can see the 12 right here. So 12 divided by 4 is?** [3] Ask a child to write the 3 in the tens place. Continue with the ones.

Help the children develop the habit of checking their answers: **Let's check our answer to see if it makes sense. Is 30 times 4 about 120?** [yes]

Before giving them a worksheet (9-21), work two more examples: 6)126 and 3)279. In the second example, if needed, help them understand that the 2 hundred 7 tens becomes 27 tens.

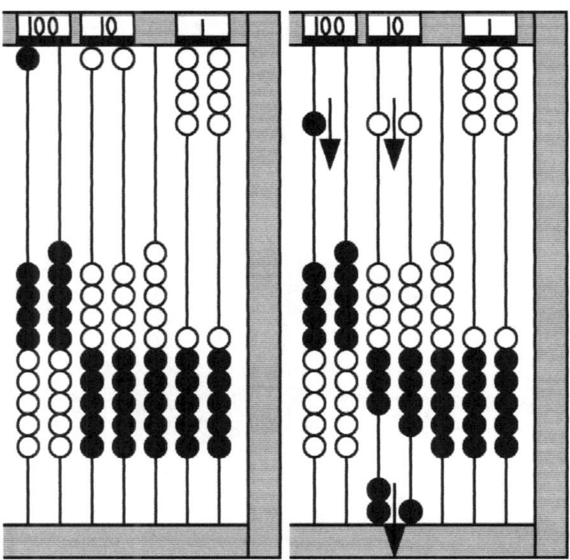

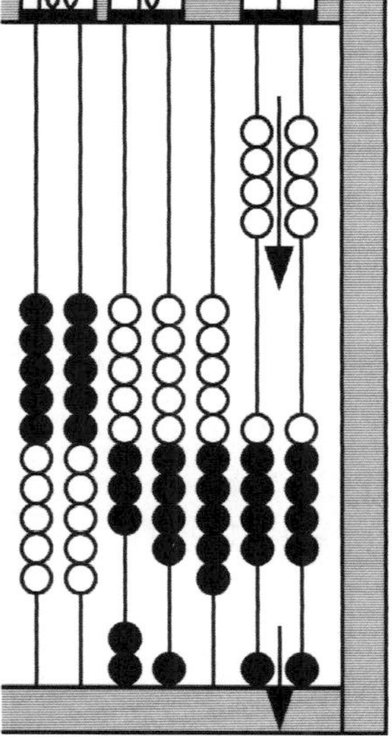

4)128	3)186	5)255
2)164	3)996	6)186
3)246	4)248	7)427
4)364	8)568	3)219
2)108	3)273	4)284
2)406	4)208	3)153

3. A common error in division is failure to write the 0s that occur in dividing numbers such 6)606. You might want to guide the children through this problem being sure they write 0s in the proper place.

For the next problem, which also has a remainder, write on the board

$$4\overline{)831}$$

Division starts as usual; the number of 4s contained in 8 hundred is 2 hundred. Write the 2. Next note that 4 is not contained in 3 tens, so write 0. Finally, the number of 4s in 31 is 7 with a remainder of 3. Enter the 7 below; remove the 31 and enter the 3 above.

Give examples 3)616 and 5)532. Then give them the worksheet (9-22).

ORAL PROBLEMS. A. Three boys decided to go on a picnic. The grocery bill came to $7.29. How much did each one owe? [$2.43]

B. In the city of Springfield, children go to school 180 days. How many weeks is that? There are only 5 days in a school week. [36 weeks]

C. A year has 365 days. How many weeks is that? There are 7 days in a week. [52 r1]

D. Mr. Black works 40 hours a week. How many hours is that in a year? [2080 hours]

Division with trading
1. Trading will not prove difficult for the children after the previous work. The important new feature is writing the remainders as superscripts (the little numbers).

Write on the board

$$3\overline{)414}$$

and enter 414 on the abacus. Ask, **How many 3s are contained in 4 hundred.** [100] Enter 1 on the hundred wire and remove 3 beads, leaving 1 as the remainder. Write the quotient in the hundreds place and ask, **How many tens are there?** [11] **To show 11 tens, we write a small 1 next to the tens place.** Write the small 1.

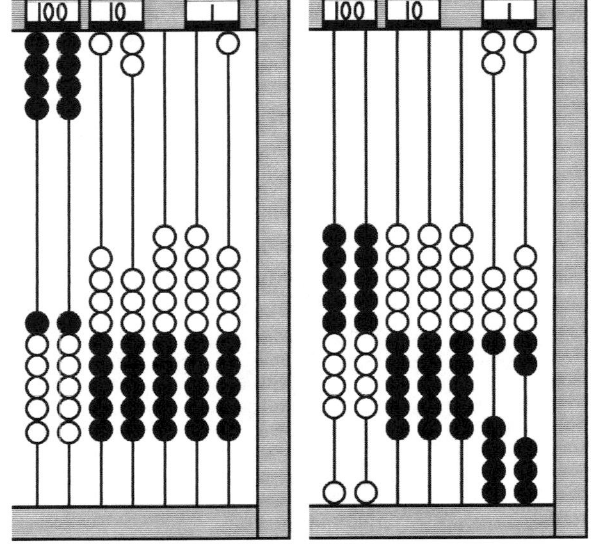

4)831	2)613	3)613
7)749	4)803	5)353
2)609	3)927	3)187
5)509	6)425	4)320
3)248	5)531	8)820
2)800	6)611	9)930

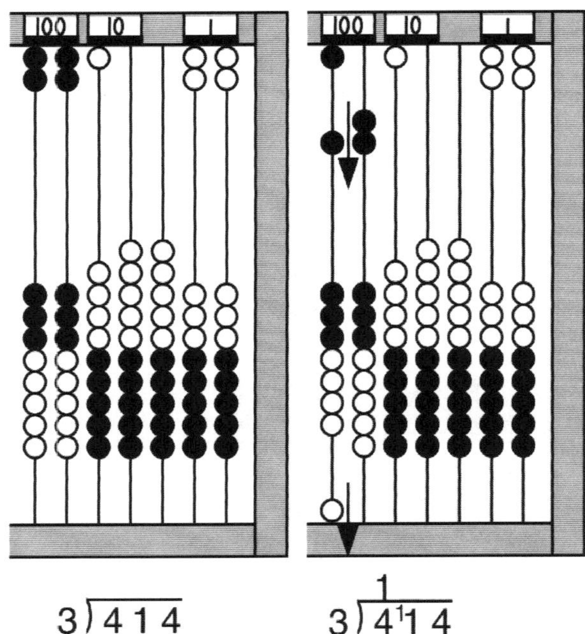

$$3\overline{)414} \qquad 3\overline{)4^{1}14}$$

Next ask, **How many 3s are contained in 11 tens?**
[3] The quotient is 3 tens and the remainder is found by
multiplying 3 x 3 and subtracting from 11. Write the 3
and the small 2. Finally the number of 3s in 24 is 8, with
no remainder.

Ask the children to work in pairs to divide 4⟌548 and
5⟌827; one child operates the abacus while the other
writes on paper. Roles are exchanged after each problem.
When the children talk aloud about what they are doing,
their understanding increases. Assign a worksheet (9-23).

ORAL PROBLEMS. Help the children solve this series of
problems by making suitable drawings, writing equations, and labeling the answers.

A. Denise makes four-legged stools. If each leg is 9 inches long, how many legs can she cut from a piece of wood
108 inches long? [12 legs.]

B. How many stools with four legs can Denise make
from each piece of wood? [3 stools]

C. If Denise buys 12 pieces of wood, how many stools
can she make? [36 stools]

D. The cost of each piece of wood is $2.95; what will the
12 pieces cost? [$35.40]

E. The sales tax for the wood amounts to $1.97. What is
Denise's total cost for the wood? [$37.37]

F. The cost for each stool totals $5.31. Denise charges
$9.00. How much does she make per stool? [$3.69]

G. How much money does Denise make if she sells all 12
stools? [$44.28]

H. Denise spends 14 minutes cutting the wood, 12 minutes assembling, and 18 minutes painting the stools. How
long does it take her to make a stool? [44 minutes] Is this
more or less than an hour? [less than an hour]

I. How long in minutes will it take her to make 12 stools?
[528 minutes]

2. After they have completed the worksheet, guide them
to divide (short division) without the abacus. Use the example at the right, 4⟌785. The number of 4s in 7 is 1 remainder 3; write the 1 above the line and the 3 before the
8. The number of 4s in 38 is 9 remainder 2; write the 9
above the line and the 2 before the 5. The number of 4s in
25 is 6 remainder 1; write the 6 and r1 above the line.

Then give them a worksheet (9-24A and B) to be done, if
possible, without the abacus. Expand the division to the
thousands with 3⟌4839 and give them similar work to do
(9-24C).

Show them how to check their work with a calculator by
multiplying the quotient by the divisor and adding the remainder. That answer should equal the dividend.

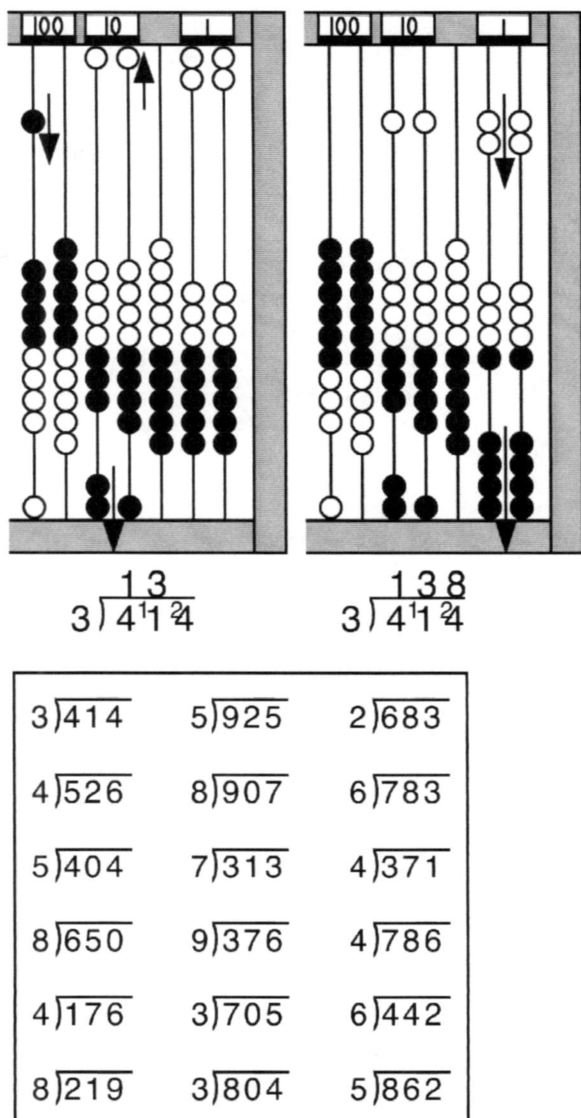

$$\frac{13}{3\,\overline{)4\,^1\!4\,^2}}\qquad\frac{138}{3\,\overline{)4\,^1\!4\,^2}}$$

3⟌414	5⟌925	2⟌683
4⟌526	8⟌907	6⟌783
5⟌404	7⟌313	4⟌371
8⟌650	9⟌376	4⟌786
4⟌176	3⟌705	6⟌442
8⟌219	3⟌804	5⟌862

$$\frac{196\,r\,1}{4\,\overline{)7\,^3\!8\,^2\!5}}$$

4⟌785	5⟌296	3⟌4839
8⟌619	6⟌123	6⟌4918
7⟌195	3⟌769	7⟌2843
2⟌137	7⟌588	4⟌9390
9⟌102	3⟌405	9⟌1321
8⟌541	7⟌340	3⟌5852

Multivides

A multivide is a self-checking exercise in multiplying and dividing. Start with any number (4 in the example) and successively multiply it by 2, then 3, and so on to 9. Then divide the final product by each number from 2 to 9. The division sign is inverted with the quotient written below. The final answer must be the same as the starting number. No remainders are possible. If one occurs, that means an error was made.

Provide the children with lined paper. Encourage them to work neatly and carefully to avoid errors. For beginners, stop at a lower number (6 in the figure) and use the division sign right side up. Gradually progress to 9.

LONG DIVISION

Children encounter three difficulties in learning long division: the algorithm itself, the special multiplication format, and guessing the correct dividend, or multiple. The algorithm is similar to that of short division if a table of multiples is used. The other two difficulties will be addressed separately.

Division with a table

The long division procedure can be done on the abacus, but the work on paper is sufficient for most children. Write on the board

$$32\overline{)9056}$$

and tell the children that now they will start to learn long division. Tell them, **We do not know the multiples of 32, so let's first construct the table of 32s.** Write in a column numbers 1 to 9. Ask the children what is 32 x 1; write 32 next to the 1. Ask them to calculate the remaining multiples. Do this by adding 32s or by employing some shortcuts. For example, the second multiple is twice the first; the third is the sum of the first and second multiples. The fourth is twice the second; the fifth is the sum of the second and the third, and so forth.

To start the division ask, **How many 32s are in 90?** Point to the table. [2] Ask, **How do we find a remainder?** [subtract] Tell them that since the numbers are larger, we will subtract neatly on paper. Write the 64 below the 90 and subtract.

Ask, **What did we do with the remainder before?** [wrote in front of the next number] **Instead of that, we will write the next number after the remainder.** Write the 5 after the 26.

Continue with, **Now we need to know how many 32s in 265.** Use the table and continue.

Leave that example on the board and give them two more

1	32
2	64
3	96
4	128
5	160
6	192
7	224
8	256
9	288

problems, 43$)\overline{2408}$ [56] and 51$)\overline{1842}$ [36 r6], to work in pairs, possibly at the board. Then give them written problems, with the outline for the multiples and some with lines for computation. (9-25 and 9-26).

Guess the multiple

The most frustrating part of long division is guessing how many times the divisor goes into the dividend, that is, guessing the correct multiple. Practice this guessing activity with the children over a period of several days.

1. Ask the children to help construct the table for 42. Alongside write the multiples of 4 and ask the children to notice the similarities. Guide them to realize that they only need to look at the 4 of the 42 and one or two digits of the multiple.

Next explain that you are going to cover up the tables and write a number. They are to guess the highest multiple. Then uncover the tables and let them see how close they are. (An overhead projector is ideal for this activity.)

Write 42$)\overline{90}$ for the first guess. [2] Also try 140 [3], 380 [9], 150 [3], and 280 [6], or any number less than 420 (10 times the divisor). Let the children work in small groups deciding on a good guess.

Repeat this activity for several days with numbers with the ones digit no higher than 3.

2. Write the multiples for 59 and the multiples for 5 and 6. Ask, **Are the 5 multiples or the 6 multiples closer to the 59 multiples?** [6] Then ask, **Why do you think this is so?** [59 is closer to 60 than to 50] Give the children numbers as above for guessing.

Use other divisors such as 38 and 69. Later include 15, 34, and 77.

"Octopus" multiplying

Until now, the children have been multiplying with all the numbers neatly aligned under each other. In long division, it is necessary to multiply with the three numbers scattered about, hence the name, "octopus."

Tell the children, **Today we will multiply in a little different way, called "octopus" multiplying. We will need it soon for long division.**

Write on the board

$$7$$
$$54$$
$$\underline{\qquad}$$

and say, **We will multiply 54 times 7 and write the product below the line; the ones will line up with the 7.** Multiply 54 x 7 and write the product as shown in the figure.

32$)\overline{9058}$

1 _____
2 _____
3 _____
4 _____
5 _____
6 _____
7 _____
8 _____
9 _____

1	42	4
2	84	8
3	126	12
4	168	16
5	210	20
6	252	24
7	294	28
8	336	32
9	378	36

1	59	5	6
2	118	10	12
3	177	15	18
4	236	20	24
5	295	25	30
6	354	30	36
7	413	35	42
8	472	40	48
9	531	45	54

$$\begin{array}{r} 3\ \text{r}69 \\ 92\overline{)345} \\ \underline{276} \\ 69 \end{array}$$

$$7$$
$$54 \quad \underline{\qquad}$$
$$378$$

Lead them through several examples, 25 x 7 and 46 x 5, before giving them the worksheet (9-27).

ORAL PROBLEMS. A. Andrea is reading a long book with 288 pages. If she can read 32 pages in an hour, how long will it take her to read the book? [9 hours]

B. Dana is making abacuses, which need 10 wires each. How many can he make with 435 wires? [43 abacuses]

54 × 7 = ___	27 × 7 = ___	47 × 9 = ___
29 × 3 = ___	26 × 8 = ___	58 × 3 = ___
83 × 6 = ___	67 × 5 = ___	72 × 3 = ___

Single-digit quotients

Now the children are ready to "solo" in some simple division problems. First give them problems with one-digit quotients. Write on the board

92)345

and ask the children to guess how many times 92 is contained in 345,that is, what is the highest multiple. [3] If necessary ask, **Does 92 go into 34?** [no] **Does 92 go into 345?** [yes] **How many times?** [3] Write the 3. Ask them to multiply and subtract.

Now ask, **What happens if we guess too high?** Rewrite the problem and use 4. Guide them to discover that they are unable to subtract. Ask, **What happens if we guess too low?** Rewrite the problem using 2. Point to the remainder and ask, **Can the remainder be more than the divisor?** [no] Tell them that everyone guesses wrong sometimes and all they need to do is erase and choose a better number.

After a few more examples, 33)97 and 73)479, give them the worksheet (9-28).

92)345	88)552	30)186
55)295	49)53	83)568
72)628	64)421	48)351
44)362	56)220	72)264
81)527	38)145	95)739

Multi-digit quotients

The final step in dividing by two-digit divisors is a combination of the previous skills. After this work, advanced children could work on dividing by three digits.

1. Tell the children that now they ready for longer numbers to divide. Write on the board

33)4634

and go through the procedure. Work 26)2045 and 48)1989 before giving them a worksheet (9-29).

33)4634	87)9823	93)7504
12)4799	39)928	28)7321

2. Now is the time to introduce the children to a small shortcut to use when 0s occur in the quotient. Write on the board

95)9679

and ask them to start as usual. The first division yields a quotient of 1 and a remainder of 1. However, 95 is not contained in 17 tens so a 0 must be written. Next we consider the ones, now a total of 179 and so forth.

```
   101
95)9679
   95
    179
```

More examples to try are 27)8251 and 18)3616 before giving them the worksheet (9-30).

95)9679	72)7794	35)7224
25)7726	16)4094	21)8135

Unit 10
Other topics

In addition to the four operations of arithmetic, there are other topics for which the AL abacus is well suited. One of these is rounding to tens. Others are finding the lowest common multiple (LCM) and the greatest common fact (GCF).

Other number bases are a natural topic for the abacus. This unit includes work with number base 4; it starts with counting and continues with the four operations. Another topic is the study of measurement. Converting from inches to feet or from ounces to pounds can be done with bead trading.

The study of squares, an extremely important topic in higher mathematics and science, is expanded to some of the common algebraic relationships. All of it is done with concrete numbers. A more everyday topic is percentages. The AL abacus with 100 beads is ideal for helping children form a general understanding of percentages.

Finally, the Al abacus can be converted into a makeshift Japanese (or Chinese) abacus by using a few strips of tape. Instructions are given for learning addition. An interesting note: in the Far East countries, abacuses are considered calculating devices; they are not used to help children understand arithmetic.

Some of these topics can be taught while the children are still learning multiplication and division.

ROUNDING

1. Tell the children that they are going to play a rounding game. Distribute only the tens from the place value cards. Enter 32 on the abacus and tell the children, **We want to change 32 into the nearest ten by moving as few beads as possible.** Ask how they would do it. Guide them to removing the 2 beads. Then ask them to show their answer with a card. [30]

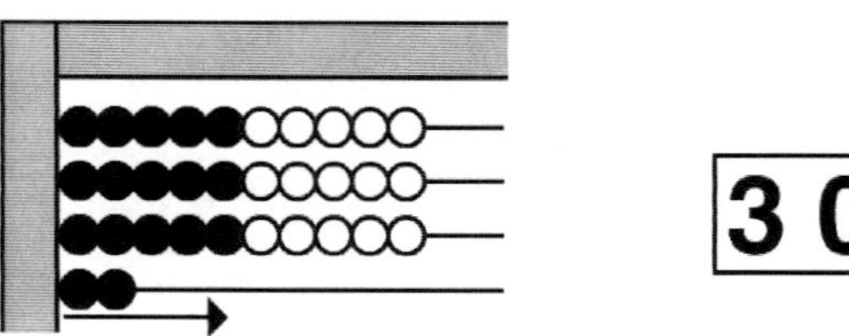

Now enter 38 and ask, **How do you change 38 into the nearest ten?** [Add 2 beads.] Ask them to show their answers. [40] Tell them, **We have ROUNDED 38 to the nearest ten, which is 40.**

Repeat for numbers 71 and 43. Then give them the worksheet (10-1A), which does not have any 5s in the ones place.

2. Numbers ending in 5 were absent from the first work

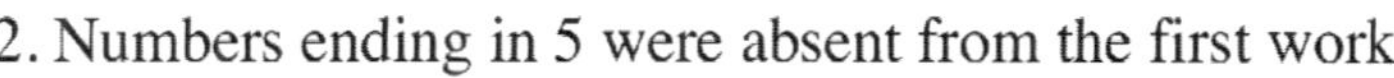

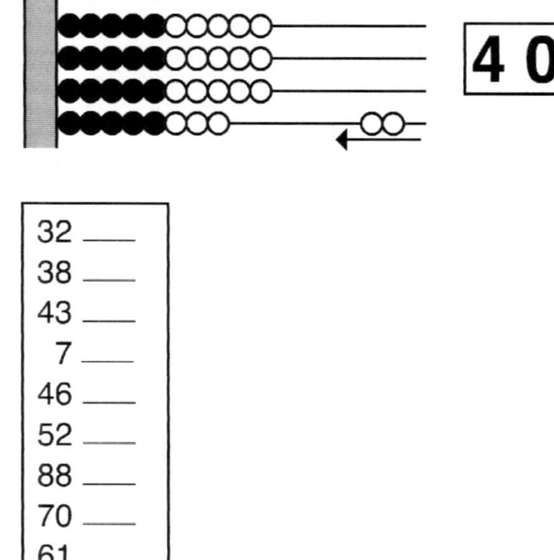

32 ——
38 ——
43 ——
7 ——
46 ——
52 ——
88 ——
70 ——
61 ——

on rounding. Enter 65 and ask the children, **How would you round 65?** Some may say 60 and some may say 70. Tell them, **Not everybody agrees, but most persons would rather enter 5 more** (round up **than remove 5, so we will call it 70.**

Repeat for 25 and 15 before giving them the next worksheet (10-1B).

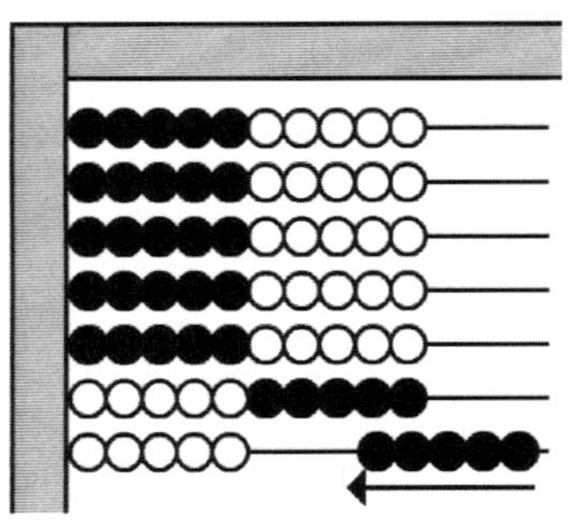

3. Distribute to the children the hundreds in the place value cards, in addition to the tens. In the vertical format, enter 257. Ask them to round to the nearest ten. To round, enter 3 beads and trade, giving the answer of 260.

Repeat for 883 [880] and 696 [700]. In the last example, after 4 beads are entered, two trades are necessary to reach 700. Then give them a practice worksheet (10-1C).

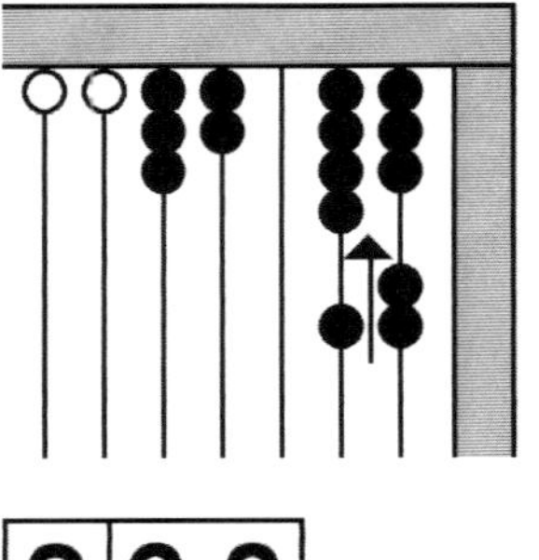

COMMON MULTIPLES

The lowest common multiple is a skill needed for fraction work. Begin by introducing the work *common*. Show to the children a picture of a dog and a picture of a cat. Draw two columns and ask them to list descriptions about each. Then ask them draw lines to descriptions that they both share. Say, **These are things they have in common.** Ask, **Is the color brown something they have in common?** [no] **Is having four legs something in common?** [yes] Repeat for other descriptions. You might want to construct a chart about two children.

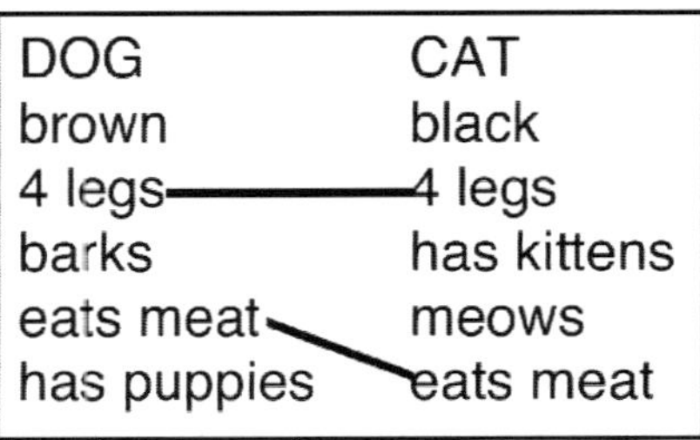

Tell the children, **Let's find some common multiples for numbers 2 and 3.** Enter two 2s on the top wires of the abacus, and enter a 3 on the lower half as shown. Ask, **Do we have a common multiple yet; are the numbers the same?** [no] Enter another 2 and another 3. Again ask, **Do we have a common multiple now?** [yes] **So 6 is a common multiple for 2 and 3.** Write it down, and continue finding the next multiples. [12 and 18]

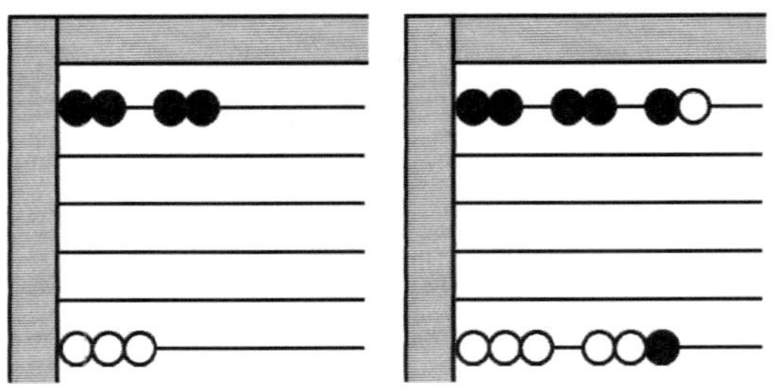

Explain that usually all we want is the first, or lowest, multiple. Ask them to find the lowest common multiple for 4 and 6 [12] and for 4 and 8 [8]. Assign them the worksheet (10-2).

Some of work can be done without the abacus. The easiest way to start is with the larger number. Decide if it is a multiple for the smaller number. If yes, the larger number is the answer. If not, continue to think of multiples of the larger number until one is found that is also a multiple of the smaller number.

65	___
19	___
85	___
96	___
88	___
30	___
55	___
49	___
83	___

257	___
345	___
552	___
136	___
125	___
489	___
563	___
372	___
605	___

2 and 3:	___
4 and 6:	___
5 and 10:	___
5 and 7:	___
2 and 10:	___
5 and 3:	___
6 and 3:	___
2 and 4:	___

PARENTHESES

Parentheses are an unnecessary source of difficulty for children when they are introduced too early. Two small baskets provide a good model for understanding the concept of parentheses.

1. In one basket, place a slip of paper with "3 + 7"; in the other, a slip with "2 + 2." Between the baskets, place a slip with a minus sign. Explain to the children, **We do what it says in the baskets first.** Ask, **How much is in the first basket?** [10] **How much is in the second basket?** [4] Ask them what the whole expression equals. [6]

Show them how to write the example: (3 + 7) - (2 + 2). Point to the parentheses and give the children the name *parentheses*. If appropriate, mention that one is called a parenthesis and more than one are called parentheses. Stress that they must always be used as a set (left and right). Repeat with other examples for the baskets, (3 x 7) + (3 x 3) and (2 + 2) x (3 + 7), calling on a child to write the equations.

Help the children develop an organized method when working with parentheses. Write on the board

$$(9 + 6) \div (1 + 2) =$$

and tell them to calculate each set of parenthesis and to write the value above it. Next, rewrite the equation without the parentheses. Finally, write the answer.

Give them more examples, such as, 10 - (3 + 4) + 11 [14] and (72 ÷ 9) + (27 ÷ 9) [11], before giving them the worksheet (10-3).

2. In this activity, the children will experience the grouping function of parentheses. They will split an array, demonstrating one form of the distributive law. One practical application is multiplying mentally, which is needed for estimating.

Enter 8 x 4 on an abacus and write on the board

$$8 \times 4 = (\quad) + (\quad)$$

and tell the children that parentheses are often used to show groups. Separate 5 x 4, using parentheses on cards, if desired, and ask them to write the two groups each in a set of parentheses. Then ask them to complete it. [8 x 4 = (5 x 4) + (3 x 4) = 20 + 12 = 32] Ask, **Is the answer what you expected?** [yes, because 8 x 4 = 32]

Next write

$$9 \times 5 = (6 \times 5) + (\quad) = \rule{1cm}{0.4pt} = \rule{0.5cm}{0.4pt}$$

and ask the children to enter the 9 x 5, separate 6 x 5, and write the remaining group. [3 x 5] Then ask them to do the calculations as above.

$$(3 + 7) - (2 + 2) = 6$$

$$\overset{15}{(9 + 6)} \div \overset{3}{(1 + 2)} = 15 \div 3 = 5$$

(9 + 6) ÷ (1 + 2) =	______	= ___
(5 + 7) - (4 + 2) =	______	= ___
9 x (3 + 7) =	______	= ___
44 ÷ (5 + 6) =	______	= ___
(18 ÷ 9) + (6 x 8) =	______	= ___
(8 x 7) - (25 x 2) =	______	= ___
(18 - 2) + (2 x 7) =	______	= ___
(3 x 5) + (3 x 5) =	______	= ___
(7 + 1) x (7 - 1) =	______	= ___

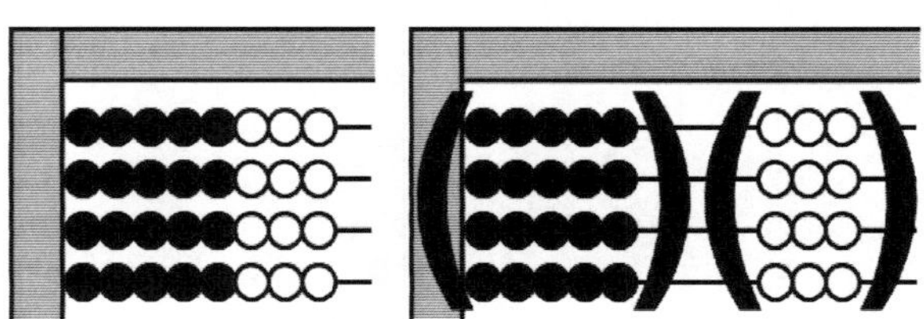

$$8 \times 4 = (5 \times 4) + (3 \times 4) = 20 + 12 = 32$$

8 x 4 = (5 x 4) + (	) =	____ =	__
7 x 8 = (5 x 8) + (	) =	____ =	__
6 x 4 = (3 x 4) + (	) =	____ =	__
8 x 8 = (5 x 8) + (	) =	____ =	__
6 x 5 = (2 x 5) + (	) =	____ =	__
5 x 5 = (4 x 5) + (	) =	____ =	__
7 x 5 = (2 x 5) + (	) =	____ =	__
9 x 5 = (6 x 5) + (	) =	____ =	__
6 x 6 = (5 x 6) + (	) =	____ =	__

Repeat for 5 x 6 = (2 x 6) + (   ). Give them a worksheet (10-4).

3. Next, the children will work with another form of the distributive law, that of combining two arrays which have one side in common. Enter 4 x 5 on an abacus and ask, **How do we write this group.** [4 x 5] Write 4 x 5 on the board. Then enter 2 x 5 on the abacus leaving a gap as shown. Again ask, **How do we write this group.** [2 x 5] Write 2 x 5. Tell them to use parentheses to keep the groups separate.

$$(4 \times 5) + (2 \times 5) =$$

Now combine the two groups. Ask, **Is the number of beads the same?** [yes] **That means we can write "=." Now how much is it?** [6 x 5]

$$(4 \times 5) + (2 \times 5) = 6 \times 5$$

If we want to write how much it is altogether, we can write another "=," and write the product.

$$(4 \times 5) + (2 \times 5) = 6 \times 5 = 30$$

Repeat for the combination of 5 x 3 and 3 x 3, and for the combination of 3 x 9 and 4 x 9 before giving them a worksheet, which consists of arrays of beads separated into two parts (10-5).

NUMBER BASE OF 4

In the same way that the study of a foreign language helps one with English, the study of other number bases helps one better understand our number base of 10. There is also a practical side, since bases of 2, 8, and especially 16, are common in computer design. This brief introduction will include the four operations, and converting to and from base 10.

Numeration

1. Tell the children that they are going to count in a different way, **We will use only the digits, 0, 1, 2, and 3. Also, we will pretend that there are only 4 beads on a wire.** Show them how to keep the four beads in the left half of the abacus. Give them the place value cards without any 4 through 9s.

First teach oral counting. Enter the first 3 beads, one a time, while saying: 1, 2, 3. Enter the 4th bead and say,

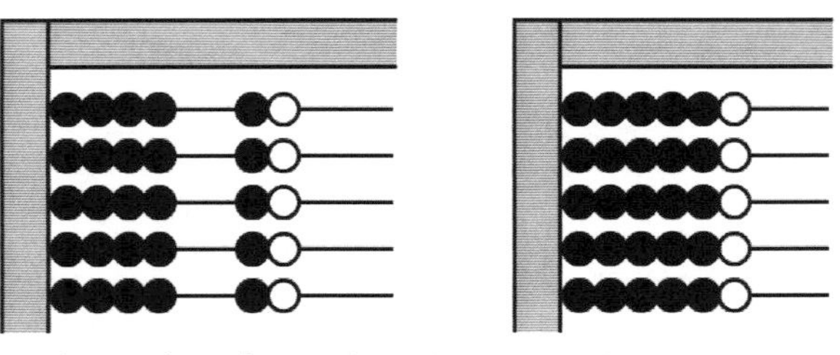

$$(4 \times 5) + (2 \times 5) = 6 \times 5 = 30$$

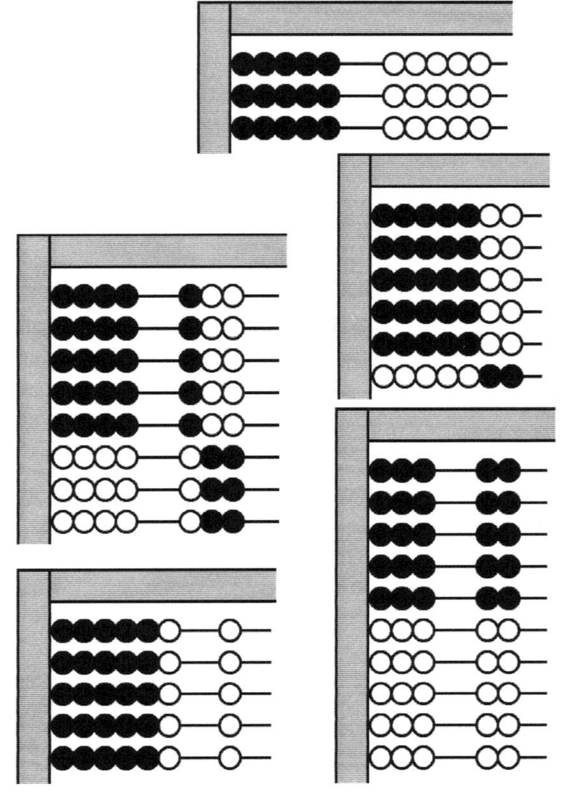

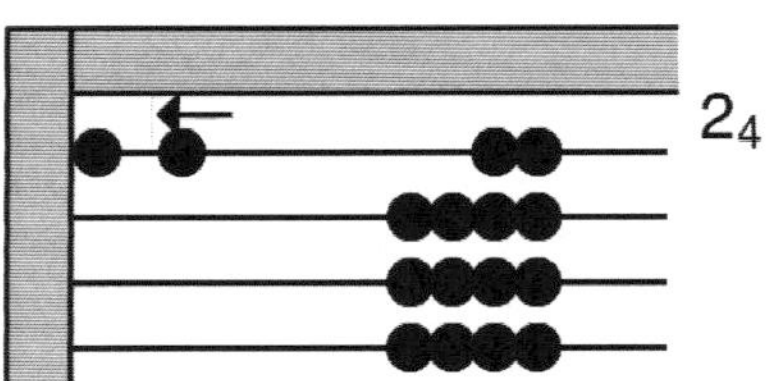

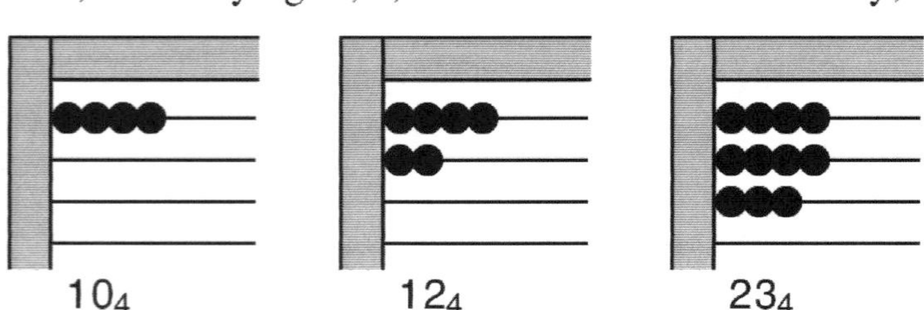

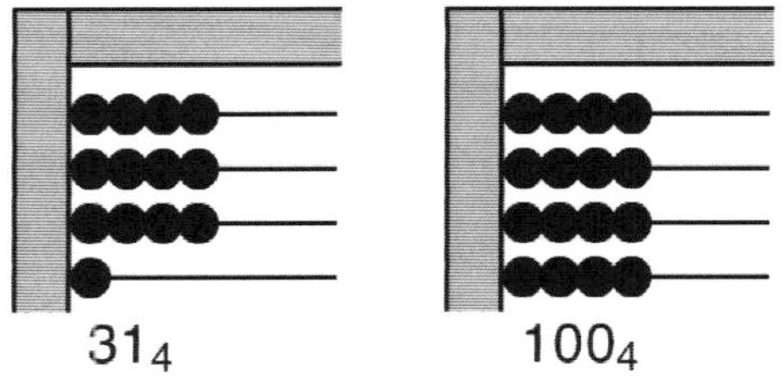

This looks like 4, but we cannot say 4. It is a whole row, so we will call it one-zero (10).

Show the place value card of 10. Continue counting, one-one, one-two, one-three, two-zero, and so forth. The fourth bead in the fourth row is one-zero-zero (100).

Show the children how to write the numbers while they count: 1_4, 2_4, 3_4, and 10_4. Explain that the little 4 means that we are using these special numbers. Have them count in pairs. Then in each pair, one person enters a quantity while the other reads it. Give the children a worksheet (10-6A) for counting from 1_4 to 100_4. Include a short line for writing each digit.

2. Turn the abacus to the vertical position and ask the children to count starting from 1. Whenever they reach 4, they must trade for a 1 in the next column. Ask them to continue counting on the worksheet (10-6B).

3. Write on the board

$$113_4 \ \underline{\quad}_4$$

and ask what number comes next. To find the next number, ask the children to enter 113 on their abacuses and to add 1. In this case, trading occurs in the ones place. So the final answer is 120_4. Also, ask them to find the next numbers for 23_4 [30_4] and for 133_4 [200_4]. Give them the worksheet (10-6C).

Adding

1. Now it is possible for the children to add in base 4 using the abacus. Write on the board

$$103_4$$
$$+\ 222_4$$

and enter the two quantities. Combine them and trade whenever there are four or more quantities in a column.

Also work with the children, 312 - 121 and 302 + 313. Then give them similar problems (10-7A).

2. For adding on paper, the Cotter Sum Line, modified for base 4, gives all the facts for addition and subtraction.

$$
\begin{array}{cccc}
0 & 1 & 2 & 3 \\
\hline
0 \ \ ^1 2 \ \ ^3 10 \ \ ^{11} 12
\end{array}
$$

Write on the board

$$313_4$$
$$+\ 213_4$$

and add as follows: 3 + 3 = 12. Write the 2 under the ones and the 1 above the tens. Next 1 + 1 + 1 = 3; write it in the tens place. Finally, 3 + 2 = 11.

1_4	31_4
2_4	32_4
3_4	33_4
10_4	100_4
11_4	101_4
12_4	102_4
13_4	103_4
20_4	110_4
21_4	111_4
22_4	112_4
23_4	113_4
30_4	120_4

Count to 100_4

$\underline{1}_4$
–
–
– –
– –
– –
– –
– –
– –
– –
– –
– –
– –
– – –

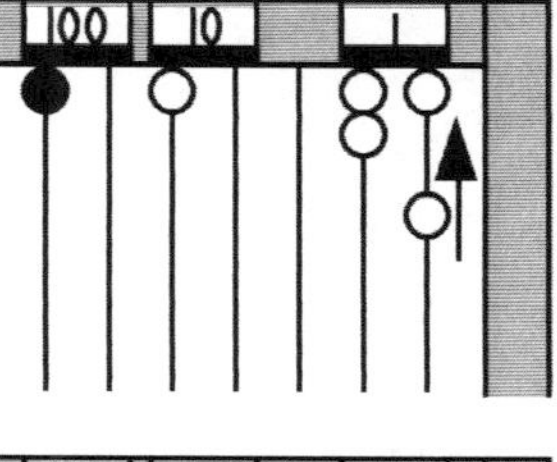

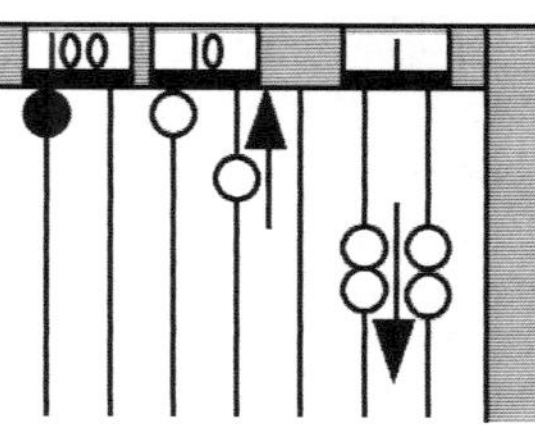

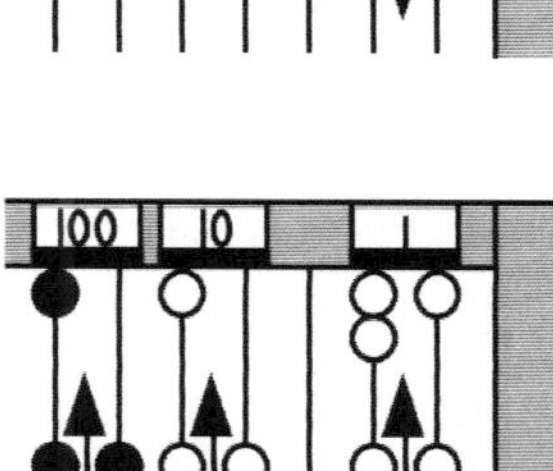

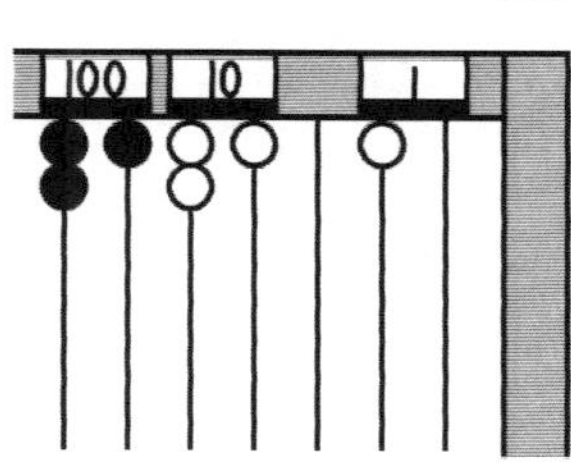

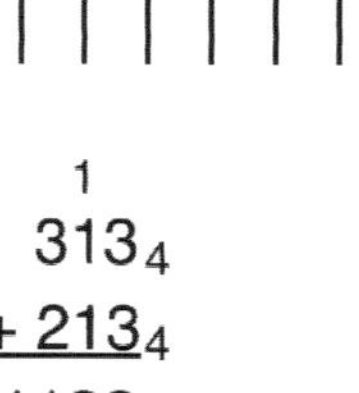

Write the next number.

113_4 ___
102_4 ___
230_4 ___
330_4 ___
103_4 ___
113_4 ___
233_4 ___
303_4 ___
333_4 ___

103_4	330_4
$+\ 222_4$	$+\ 103_4$
203_4	332_4
$+\ 331_4$	$+\ 302_4$
321_4	1320_4
$-\ 202_4$	$-\ 322_4$
2031_4	3201_4
$-\ 1302_4$	$-\ 2302_4$

$$\overset{\ \ 1}{313_4}$$
$$+\ 213_4$$
$$\overline{1132_4}$$

Give them more problems to add and subtract (10-7B).

THOUGHT QUESTION. Do the even and odd rules work the same in base 4? [yes]

Multiplication and division

1. Draw a blank 4 x 4 grid and help the children build the multiplication table. With it they will be able to multiply and divide in base 4. Write on the board

$$132_4$$
$$\times\ 31_4$$

and guide them in multiplying as shown. Let them try multiplying $23_4 \times 21_4$ and $123_4 \times 13_4$ before giving them work (10-8A) on their own.

2. Division is not difficult with the aid of the multiplication table. First is a problem for short division. Write on the board

$$3\overline{)2213}$$

and guide the children in the dividing as shown. Give them $2\overline{)1312}$ [323] and $3\overline{)1130}$ [132 r2] as examples.

Next is a long division problem. Write on the board

$$12_4\overline{)3211}_4$$

and help them in dividing. Two other examples to give them before the worksheet (10-8B) are $13\overline{)3213}$ [201] and $31\overline{)20312}$ [223 r13].

Converting between base 4 and base 10

1. The easier conversion is from base 4 to base 10. The first digit on the right is the ones in any base. In base 4, the value of the second digit from the right is 4 times the digit. The value of the third digit is 4 times 4, or 16, times the digit. The value of the fourth is 4 x 16, or 64, times the digit, and so forth.

To convert a number to base 10 calls for multiplying each digit by the corresponding value. See the example shown. Give the children calculators to speed the work. Help the children organize their work.

Two other examples to try are: 2113_4 [150] and 1203_4 [99]. Give the children the worksheet (10-9A).

2. To convert from base 10 to base 4, use the place value equivalents. Subtract the highest quantities and write it in the correct column. The children will discover the rules as they work from the simpler to harder problems. It is the reverse process of converting from base 4. Other algorithms exist which simplify the calculations, but elementary children would find the mathematics too difficult.

Write on the board

$$16_{10}$$

$\times_4$	1	2	3
1			
2			
3			

$\times_4$	1	2	3
1	1	2	3
2	2	10	12
3	3	12	21

$$132_4$$
$$\times\ 31_4$$
$$132$$
$$\underline{11220}$$
$$12012_4$$

$$132_4 \qquad 123_4$$
$$\times\ 31_4 \qquad \times 23_4$$

$$213_4 \qquad 302_4$$
$$\times\ 33_4 \qquad \times 30_4$$

$$313\ r2$$
$$3\overline{)22^{1}1^{2}3}$$

$$212\ r1$$
$$12\overline{)3211}$$
$$\underline{30}$$
$$21$$
$$\underline{12}$$
$$31$$
$$\underline{30}$$
$$1$$

$$2\overline{)3221} \quad 13\overline{)3213}$$

$$3\overline{)3001} \quad 23\overline{)3102}$$

$$2\overline{)1203} \quad 11\overline{)2103}$$

256	64	16	4	1
	3	2	1	2_4

$$3212_4 = (64 \times 3) + (16 \times 2) + (4 \times 1) + 1$$
$$= 192 + 32 + 4 + 1 = 229_{10}$$

256	64	16	4	0
		1	0	0_4

$3212_4 = $ _____ $_{10}$		$16_{10} = $ _____ $_4$
$1021_4 = $ _____ $_{10}$		$32_{10} = $ _____ $_4$
$320_4 = $ _____ $_{10}$		$33_{10} = $ _____ $_4$
$2003_4 = $ _____ $_{10}$		$65_{10} = $ _____ $_4$
$333_4 = $ _____ $_{10}$		$128_{10} = $ _____ $_4$
$2031_4 = $ _____ $_{10}$		$200_{10} = $ _____ $_4$

and ask the children how they would change it to base 4.
Since 16 is the value of the third place, the solution is
100_4. Similarly, 36 has two 16s and a 4, so $36_{10} = 210_4$.

Give the children other problems (10-9B).

Other bases

Any of the other number bases can be approached in the
same manner. See worksheets 10-10 through 10-13 for
work in base 8.

FACTORS

Children sometimes confuse lowest common multiple
(LCM) and greatest common factor (GCF), so it is wise
to separate the topics.

1. Review the meaning of factor: in 2 x 3 = 6, 2 and 3 are
factors. It is also a number by which a number can be divided without a remainder. Ask the children to enter 16 on
their abacuses and to find pairs of factors that multiplied
together equal 16. At first, let them find them in random
fashion.

Later, help them to become organized by starting with 1.
Ask, **Is 1 a FACTOR?** [yes] **What is the other factor?** [16] Write them down and continue, **Is 2 a factor?**
[yes] **What is the other factor?** [8] Write it down.
Three is not a factor because it has a remainder of 1. Four
is a factor, but the other factor is also 4. There is no need
to go farther, but let the children discover this. So the factors of 16 are: 1, 16, 2, 8, 4. Let them find the factors for
15 [1, 15, 3, 5] and 18 [1, 18, 2, 9, 3, 6] before giving
them the worksheet (10-14).

2. In reducing fractions, it is necessary to find the
GREATEST COMMON FACTOR. Ask the children to
find the factors for 18 and 12, and then to choose the
greatest common factor. Give them a worksheet with the
same type of problems (10-15).

16: ______	12 and 18: __
12: ______	24 and 6: __
10: ______	15 and 10: __
6: ______	21 and 14: __
20: ______	2 and 10: __
8: ______	18 and 15: __
7: ______	16 and 24: __
9: ______	8 and 10: __
13: ______	12 and 9: __

MEASUREMENT

1. An activity similar to bead trading is valuable for
hands-on conversion of units, especially those in the U.S.
Customary system. Give the children 12-inch rulers.
Show them an *inch* on the ruler and show them how to
measure in inches. Let them find small objects in the room
to measure.

Now take two pieces of paper, one 7 inches wide and the
other 2 inches wide. Ask, **What will we get if we lay
the sheets of paper side by side and measure
them?** [9 in] Write on the board

$$\begin{array}{r} 7\text{ in} \\ +\,2\text{ in} \\ \hline 9\text{ in} \end{array}$$

and explain that this is how we write it, **The *in* means inches; we do not need to write the whole word.**

Hold up the ruler and tell them it measures one *foot*, so-called because it was the length of a king's foot. Have groups of children measure objects more than a foot long. Show them how to use several rulers end-to-end for measuring.

Next write on the board

$$7 \text{ in}$$
$$+\ 9 \text{ in}$$

and ask the children to add it. Ask, **How many inches in a foot?** [12] **So we need to trade 12 inches for 1 foot.** Show them how to use the abacus for the trading. The right two wires represent inches; the fourth and fifth wires from the right represent feet.

This time give them a problem with feet and inches

$$2 \text{ ft } 8 \text{ in}$$
$$+\ 1 \text{ ft } 6 \text{ in}$$

to solve in the same way. After another example, 3 ft 9 in + 5 ft 11 in, give them a worksheet (10-16A).

2. Measurement problems involving subtraction can also be done initially on the abacus. Write on the board

$$4 \text{ ft } 4 \text{ in}$$
$$-\ 2 \text{ ft } 6 \text{ in}$$

and ask the children to subtract. They first subtract 2 from 4, but write 1, noting that 6 inches is greater than 4 inches. Trading results in 12 inches being added to the 4, giving 16 inches. The 16 should be written above the 4. See the figure.

Work with them on the next examples, 3 ft 8 in - 2 ft 9 in and 10 ft - 3 ft 1 in. Assign a worksheet (10-16B).

3. Multiplication problems involving units are solved in the same way as the addition problems (10-17A).

For problems involving yards and feet, show them a yardstick with three rulers lined up below it. Let them work the problems with the objects in view. Assign a worksheet (10-17B).

4. Provide them with problems (10-18) concerning pounds and ounces, and also with pounds and tons. The latter problems cannot easily be done on the abacus; however, most children will no longer need it.

For problems (10-19) involving cups, quarts, and gallons, bring appropriate containers. Let the children verify the relationships by using water.

Also give them problems using time: weeks and days, and years and months (10-20). Finally, give them work with hours and minutes and seconds (10-21).

$$7 \text{ in}$$
$$\underline{9 \text{ in}}$$
$$16 \text{ in } = 1 \text{ ft } 4 \text{ in}$$

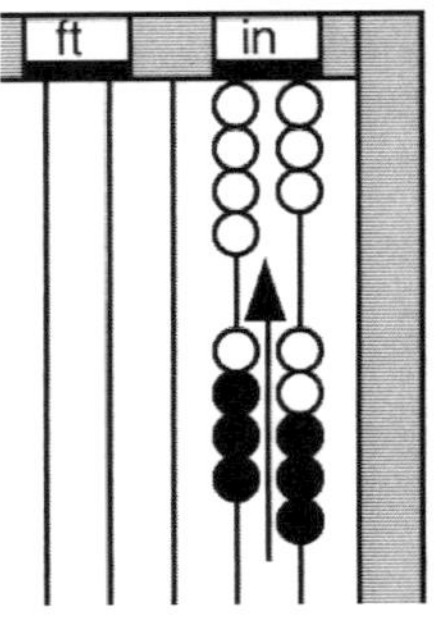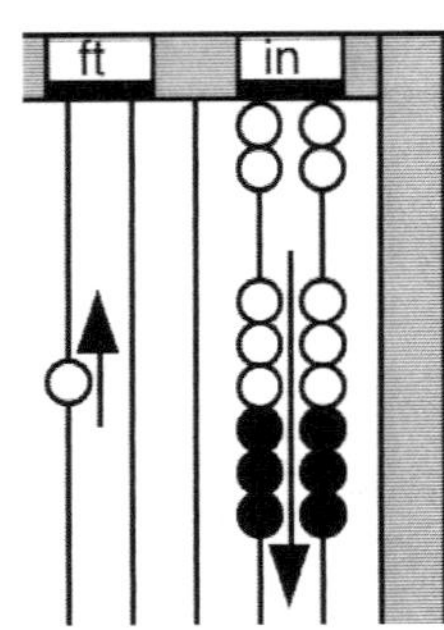

2 ft	8 in
+ 1 ft	6 in

3 ft 14 in = 4 ft 2 in

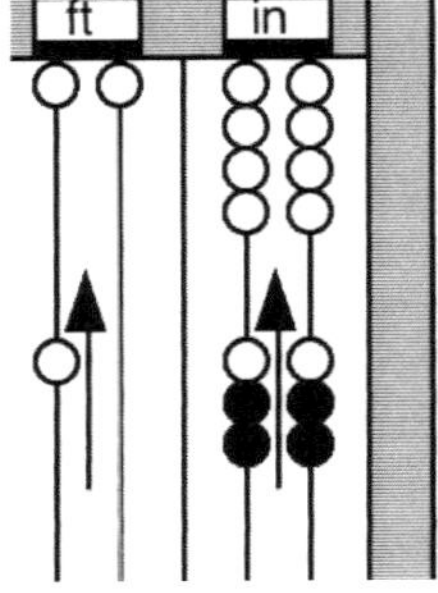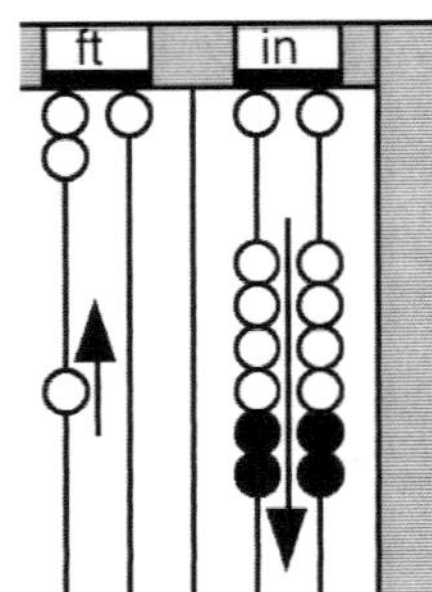

2 ft	8 in
+ 1 ft	6 in
3 ft	9 in
+ 4 ft	8 in
5 ft	11 in
+ 2 ft	10 in

$$\overset{16}{\phantom{4 \text{ ft}}}$$

4 ft	4 in
- 2 ft	6 in
1 ft	10 in

4 ft	4 in
- 2 ft	6 in
7 ft	3 in
- 4 ft	8 in
5 ft	0 in
- 2 ft	9 in

ORAL PROBLEMS. Benjamin practiced his clarinet last week for 2 hrs 45 min. This week he practiced 1 hr 50 min. How much time did he practice during the two weeks? [4 hrs 35 min]

ON TO THE MILLIONS

Children frequently hear large numbers. They can be introduced to them as soon as they thoroughly understand 1, 10, 100, and 1000 and how to write them.

Constructing larger numbers

Use the Base 10 Picture Cards with the children to review and expand the number system.

Start with the 1-cube and say, **This is a one; what is 10 ones?** [10-bar] **What is 10 tens?** [hundred-square] **What is 10 hundreds?** [thousand-cube] If available, call upon different children to build the quantities. Also ask them to match the place value cards with the quantities.

Point out that the thousand-cube looks like the one-cube, but it is 1000 times greater. Take the cube and ask the children to imagine ten of them laid out in a straight line. **What would we have?** [ten-thousand-bar] **Now imagine ten of those bars laid side by side. What would that be?** [hundred-thousand-square] **Stack ten of those squares and we would have the *million*-cube—a cube that is like the thousand-cube, but one thousand times greater.**

Use the place value cards to show them how they are written. Explain that the comma separates the thousands.

Then give them numbers to construct with the place value cards and to enter on the abacus. Start by writing on the board

32,749.

On the abacus, use single wires to represent each place value. Use Side 2 (with the numbers at the bottom) and tape commas as shown. Skip the two right most wires, which will enable the thousands to begin with the color change. The two right wires can be used for showing the cents for dollar amounts. Finally, ask one of the children to read the quantity. Remind them that when they see a comma, they are to say a word.

Let the children practice with 152,893 and 890,007 and 1,400,357. The word "and" must not be said when reading these numbers; it is reserved to indicate the decimal point in a number, such as 13.2 (13 and 2 tenths). Give them the worksheet (10-22).

Also give them problems in the four operations with these higher numbers (10-23 and 10-24).

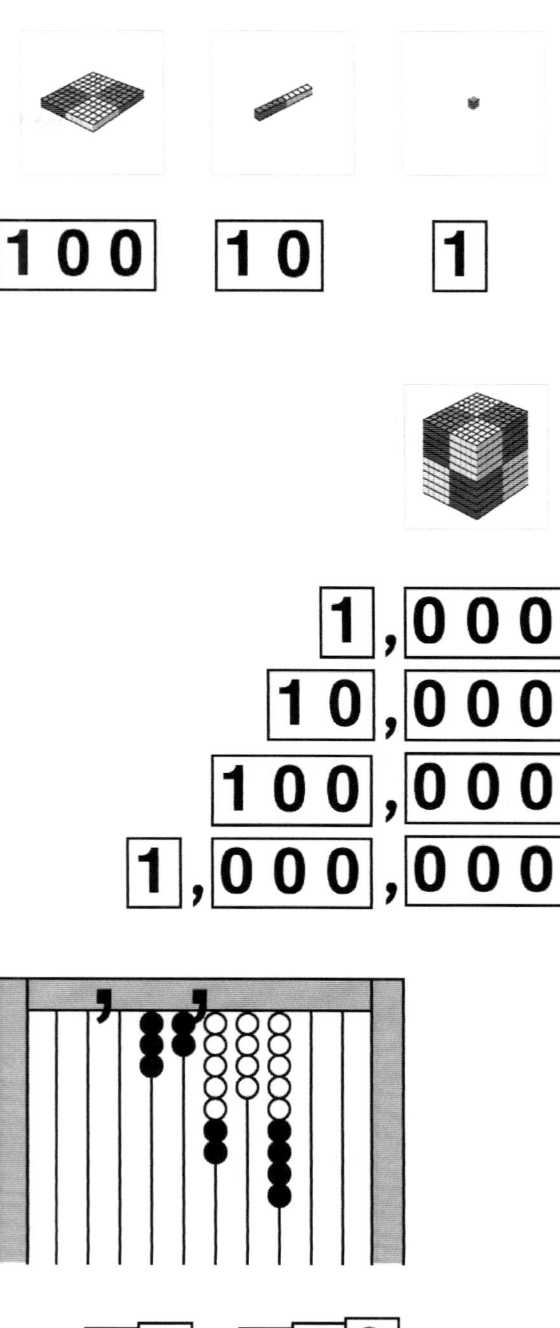

SQUARES

The study of squares is interesting in its own right. It is also needed for the study of area and is used extensively in higher mathematics. A number squared is the product of a number multiplied by itself.

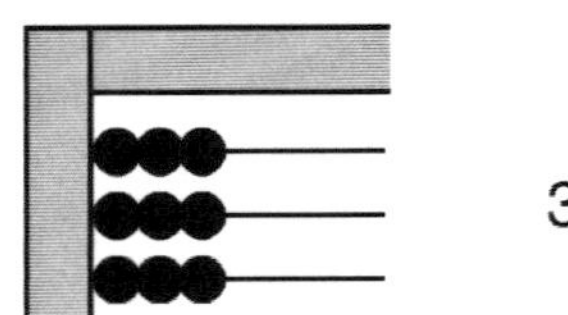

Introduction

Tell the children that today are going to learn a different way to write something they already know. Write on the board

$$3^2 =$$

and tell the children that the little 2 means that the number 3 is SQUARED. It means we have 3 two ways: 3 rows of 3s, or 3 x 3. Enter it on the abacus. Ask, **How much is 3 squared?** [9] Repeat for 6^2 [36], 10^2 [100], and 1^2 [1]. Children sometimes want to say that 1^2 is 2; remind them that it means 1 x 1.

Next ask the children to do the reverse. Write

$$81 = \_\_^2$$

and ask what number multiplied by itself equals 81 [9]. Refer to a multiplication table and ask where the squares can be found. [along the diagonal from upper left to lower right]

Now ask them to find the square for a harder number, 121. Tell them to guess and then check their answer [11]. Give them written work (10-25) for squaring numbers and finding squares.

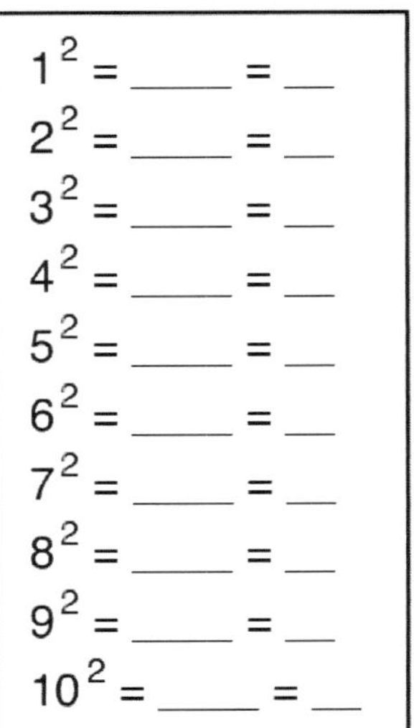

Building squares

1. Another way of calculating squares is to start with 0 and add in order the odd numbers. Each sum gives the next square. Give the children a prepared worksheet (10-26A) as shown.

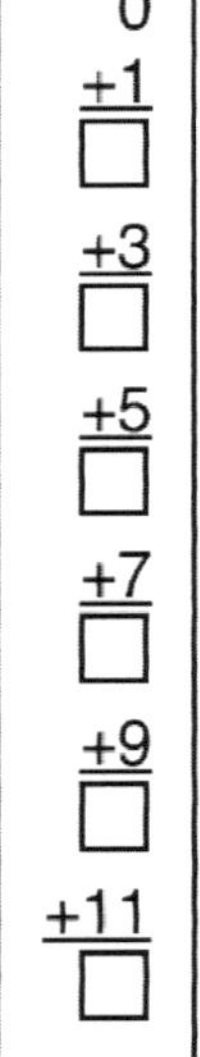

2. Play a counting game with two teams of children. Teams alternate saying a number in turn from 1 to 100. However, if the number is a square, the word "square" must be said instead of the number; a person forgetting is out of the game. After 100 is reached, start again with 1. The team with the most members at the end of two rounds to 100 is the winner.

An interesting result occurs when adding consecutive odd numbers. Write on the board

$$1 = \_\_ = \_^2 \, [1^2]$$
$$1 + 3 = \_\_ = \_^2 \, [2^2]$$
$$1 + 3 + 5 = \_\_ = \_^2 \, [3^2]$$
$$1 + 3 + 5 + 7 = \_\_ = \_^2 \, [4^2]$$
$$1 + 3 + 5 + 7 + 9 = \_\_ = \_^2 \, [5^2]$$

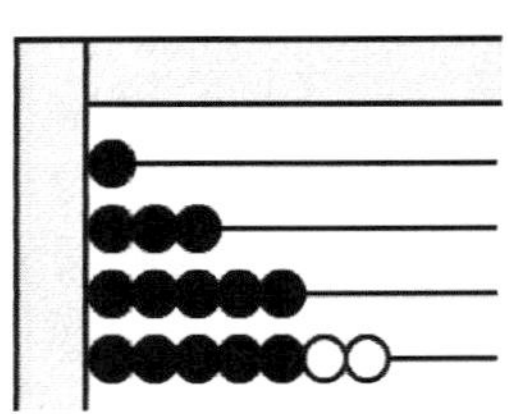
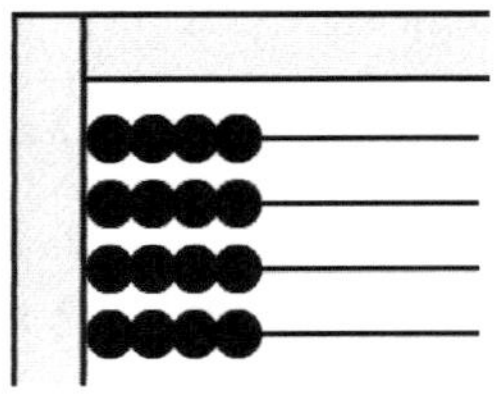

and assign a small group of children to each problem. Ask them to enter each addend on a separate line, then to use take and give to change the quantity into a square. The figures show the work for the fourth problem. Tell the children to write their answers. Ask them what pattern they see. Guide them to discover that the sum is the square of the number of addends. For example, the third row has 3 addends and its sum is 9, or 3^2.

As an application, give them longer sums, such as, $1 + 3 + 5 + 7 + 9 + 11 + 13 + 15 + 17 + 19$. [100] Or, ask them to write the sum that would total 64. [$1 + 3 + 5 + 7 + 9 + 11 + 13 + 15$]

3. This next activity demonstrates that if two numbers that differ by 2 are multiplied and 1 is added, the results will be the square of the middle of the two numbers. Tell the children to look for an interesting result.

Write on the board

$$5 \times 3 + 1 =$$

and ask them to enter it on their abacuses. Next ask them to change it into a square. See the figure. Repeat for $4 \times 2 + 1$ and $8 \times 10 + 1$. Suggest they look for the pattern with the easy problems on the worksheet (10-26), so they can do the harder ones without the abacus.

This results can be seen on a multiplication table. The diagonals adjacent to the diagonal with the squares are one less than the square.

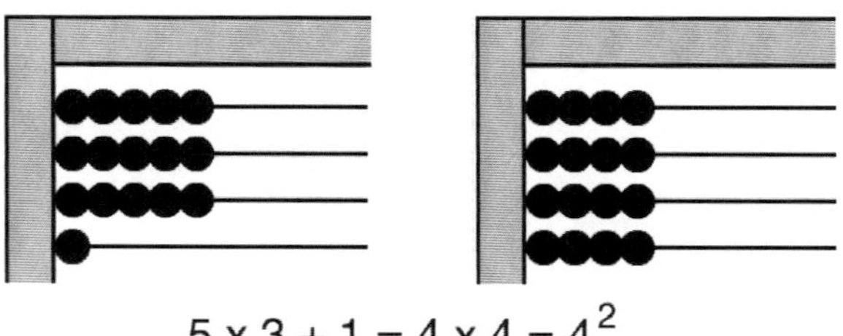

$$5 \times 3 + 1 = 4 \times 4 = 4^2$$

$$
\begin{aligned}
5 \times 3 + 1 &= \underline{\hspace{1cm}} = \underline{\hspace{0.5cm}} \\
6 \times 4 + 1 &= \underline{\hspace{1cm}} = \underline{\hspace{0.5cm}} \\
10 \times 8 + 1 &= \underline{\hspace{1cm}} = \underline{\hspace{0.5cm}} \\
2 \times 4 + 1 &= \underline{\hspace{1cm}} = \underline{\hspace{0.5cm}} \\
7 \times 9 + 1 &= \underline{\hspace{1cm}} = \underline{\hspace{0.5cm}} \\
6 \times 8 + 1 &= \underline{\hspace{1cm}} = \underline{\hspace{0.5cm}} \\
7 \times 5 + 1 &= \underline{\hspace{1cm}} = \underline{\hspace{0.5cm}} \\
9 \times 11 + 1 &= \underline{\hspace{1cm}} = \underline{\hspace{0.5cm}}
\end{aligned}
$$

Finding the next square

A simple method exists to find a square of a number if the square of the preceding number is known. Start with 4^2 and use it to find 5^2. Write on the board

$$\underline{\hspace{0.5cm}}^2 = 4^2 + 4 + 5 = \underline{\hspace{0.5cm}}$$

and show the children how to enter the quantities as shown. Ask them what new square is formed. [5^2] Repeat for $8^2 + 8 + 9$ [9^2 is the new square formed] and $9^2 + 9 + 10$. [10^2 is formed]

Once the pattern is understood, the children can use it to find squares of higher numbers. Give them calculators for checking their work (10-27).

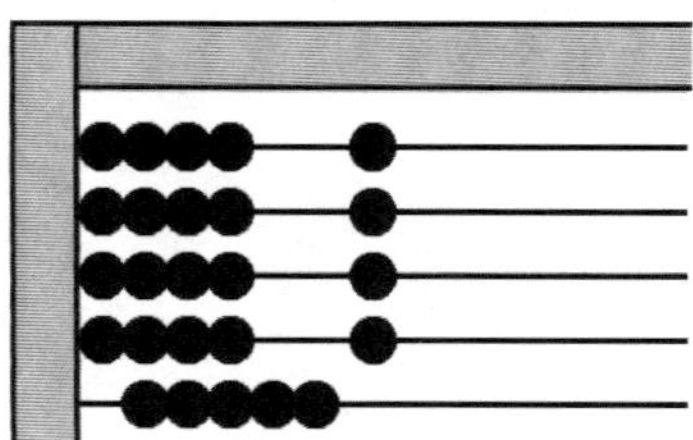

$$5^2 = 4^2 + 4 + 5 = 25$$

$$
\begin{aligned}
\underline{\hspace{0.5cm}}^2 &= 4^2 + 4 + 5 = \underline{\hspace{0.5cm}} \\
\underline{\hspace{0.5cm}}^2 &= 2^2 + 2 + 3 = \underline{\hspace{0.5cm}} \\
\underline{\hspace{0.5cm}}^2 &= 5^2 + 5 + \underline{\hspace{0.5cm}} = \underline{\hspace{0.5cm}} \\
\underline{\hspace{0.5cm}}^2 &= 3^2 + 3 + \underline{\hspace{0.5cm}} = \underline{\hspace{0.5cm}} \\
11^2 &= \underline{\hspace{0.5cm}} + \underline{\hspace{0.5cm}} + \underline{\hspace{0.5cm}} = \underline{\hspace{0.5cm}} \\
12^2 &= \underline{\hspace{0.5cm}} + \underline{\hspace{0.5cm}} + \underline{\hspace{0.5cm}} = \underline{\hspace{0.5cm}} \\
13^2 &= \underline{\hspace{0.5cm}} + \underline{\hspace{0.5cm}} + \underline{\hspace{0.5cm}} = \underline{\hspace{0.5cm}} \\
14^2 &= \underline{\hspace{0.5cm}} + \underline{\hspace{0.5cm}} + \underline{\hspace{0.5cm}} = \underline{\hspace{0.5cm}} \\
15^2 &= \underline{\hspace{0.5cm}} + \underline{\hspace{0.5cm}} + \underline{\hspace{0.5cm}} = \underline{\hspace{0.5cm}}
\end{aligned}
$$

Squaring a sum

The abacus clearly shows the results of a quantity like $(5 + 3)^2$. The 5 and 3 are both squared, but there are also two other products, actually identical. In this example, it is 5 x 3. Write on the board

$$(5 + 3)^2$$

and a child to enter it on an abacus. Remind them that 5 + 3 must be in both directions: the number of rows and the number of columns. Separate 5^2 and 3^2. Help them write the results as shown in the figure.

Give them $(4 + 1)^2$ and $(5 + 2^2)$ to work before giving them the worksheet (10-28).

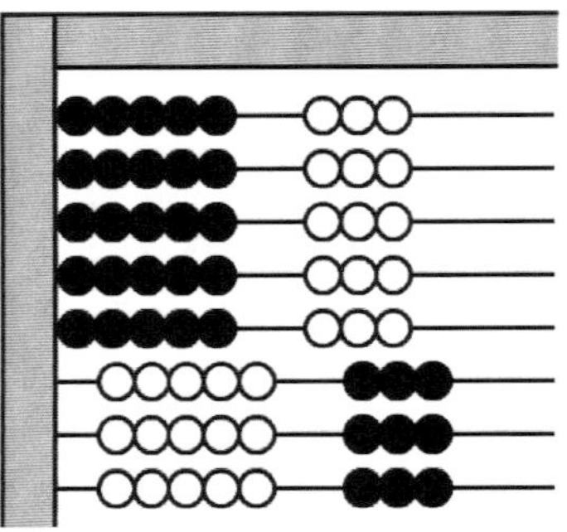

$$(5 + 3)^2 = 5^2 + 2 \times (5 \times 3) + 3^2$$
$$= 25 + 30 + 9 = \underline{64}$$

$$(5 + 3)^2 = 5^2 + 2 \times (5 + 3) + 3^2$$
$$= \underline{\qquad} = \underline{\quad}$$

$$(3 + 2)^2 = \underline{\ }^2 + 2 \times (3 + 2) + \underline{\ }^2$$
$$= \underline{\qquad} = \underline{\quad}$$

$$(4 + 3)^2 = \underline{\ }^2 + 2 \times (\underline{\ } + \underline{\ }) + \underline{\ }^2$$
$$= \underline{\qquad} = \underline{\quad}$$

$$(10 + 2)^2 = \underline{\qquad\qquad}$$
$$= \underline{\qquad} = \underline{\quad}$$

Dividing a square by a square

Calculating how many small squares will fit in a bigger square is a problem that arises in the study of area. Write on the board

$$\frac{6^2}{3^2}$$

and ask the children to guess the answer. Then ask them to use the abacus to calculate the answer. Later, for work without the abacus, the easiest way to solve these problems is to divide and square the quotient. Give them the worksheet (10-29).

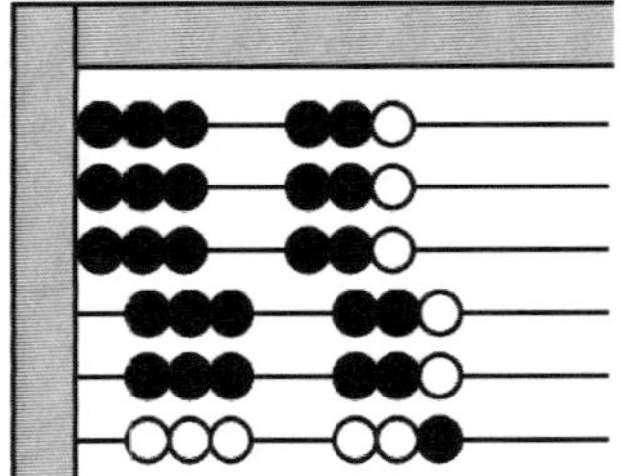

$$\frac{6^2}{3^2} = \underline{\quad}$$
$$\frac{8^2}{4^2} = \underline{\quad}$$
$$\frac{10^2}{2^2} = \underline{\quad}$$
$$\frac{9^2}{3^2} = \underline{\quad}$$
$$\frac{4^2}{2^2} = \underline{\quad}$$
$$\frac{10^2}{5^2} = \underline{\quad}$$
$$\frac{6^2}{2^2} = \underline{\quad}$$

PERCENTAGE

Percentages are a common way to express relationships; children hear them mentioned frequently. They can easily understand the concept with the 100 beads on the abacus.

A fraction of a number

Before finding the common percentages, the children need to understand how to find a fraction of a number. To introduce children to fractions, see the *Math Card Games* chapter on fractions.

Review that 6/3 is 6 divided by 3, which equals 2, and 3/3 equals 1. Also, 1/3 is 1 divided by 3, or one-third. The fraction 2/3 can be thought of as two one-thirds. Write on the board

$$\tfrac{2}{3} \text{ of } 12$$

explain that it means 12 is to divided into 3 groups and we take 2 of those groups.

Ask the children to enter 12 and to divide it into 3 groups, that is, on 3 wires. Ask, **How many are in each group?** [4] **How many are in 2 groups?** [8] **So, 2/3 of 12 is 8.**

Repeat for 1/4 of 16 and 5/6 of 12. Give them a worksheet (10-30).

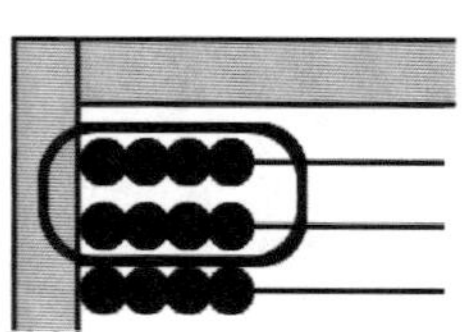

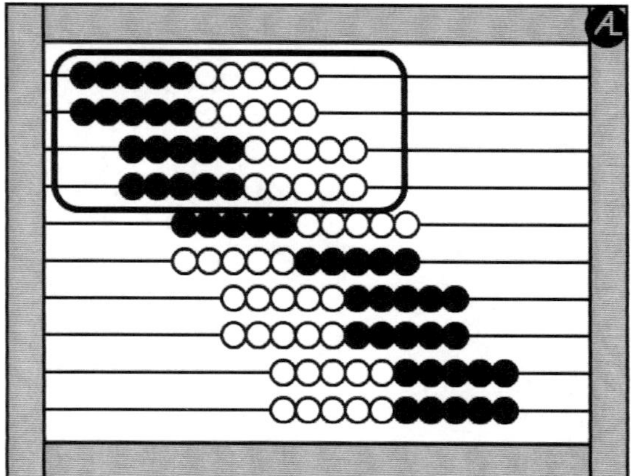

A fraction of 100

Tell the children that now they will learn about percents. A fraction of the special number, 100, is called a percent. Start by writing on the board

$$\tfrac{2}{5} \times 100\% = \underline{\quad}$$

Hold up an abacus and say, **All the beads on this whole abacus is 100%. So what percent would 2/5 be?** Guide the children in dividing the abacus into fifths [20] and taking 2 of the fifths [40]. **Two-fifths is 40%.** Ask a child to write the result.

Let the children find 1/2 as a percent. [50%] Ask the children to find 1/4 as a percent. [25%] They may need help in dividing the abacus into fourths. The easiest way is to separate by color; see the figure. To find 1/8, they will need to take one-half of 1/4, which will be 12½%.

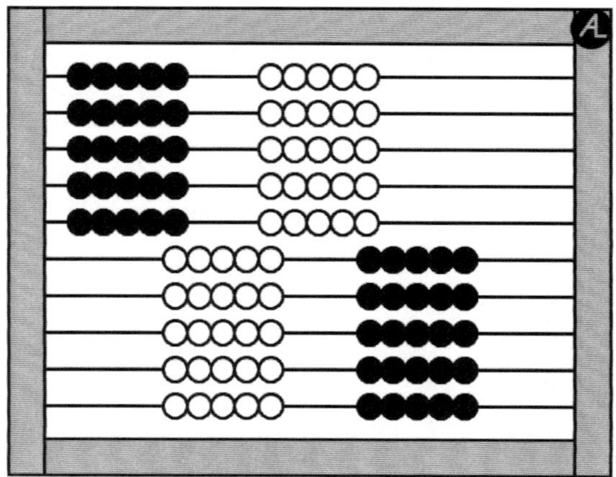

If the children are interested, help them find 1/3 as a percent. Ninety can be divided into thirds with 10 left over. Ten can also be divided into thirds, with 1 left over. So 1/3 is 33-1/3%.

On some calculators, the percent can be found by dividing the numbers and pressing the % key. To find 2/5, for example, the key sequence is: 2, ÷, 5, and %.

Give the children a worksheet for changing fractions into percents (10-31).

THOUGHT QUESTION. If one fraction is larger than another fraction, will the corresponding percents be larger? [yes]

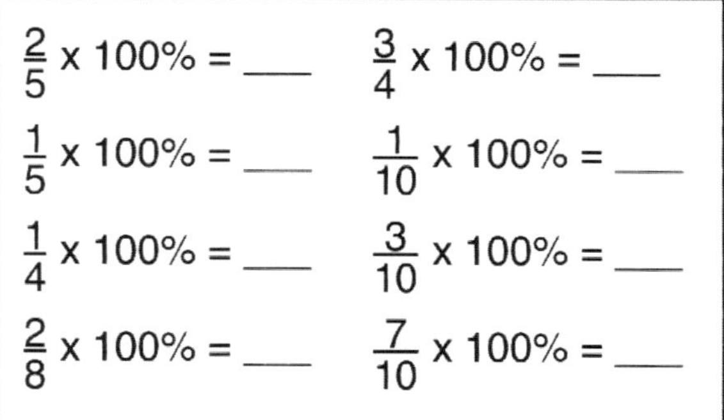

AVERAGE

Pose this problem to the children, **Danielle has 7 flowers; Casey has 8; and Keith has 3. How can they divide them so all have the same number?** Ask them to enter the quantities on their abacuses. Let the children come up with various solutions. Two possible methods are to use take and give or to add them together and divide by 3.

Tell them that the answer is called the *average*. If the flowers were divided evenly, each person would have 6.

Ask them to find the average of 5 and 9. [7] Guide them to recognize the average of two numbers as being the same as the middle of the numbers.

Next give them larger numbers to average. **If Derek had scores of 35, 29, 87, and 53 following a game, what is his average?** Since these numbers are too large to do by take and give, suggest that the children add them together and divide. Ask, **What number do we divide by?** [4] Before doing the arithmetic, ask the children to guess the answer. Record their responses for comparison later. Encourage neat work using the format shown in the figure.

Ask them what the answer means. One interpretation is that if Derek had scored 51 in each game, he would have a total of 204 points.

Give them the worksheet (10-32).

THOUGHT QUESTIONS. After they have completed the worksheet, ask the children the following questions.

A. Can an average ever be the same as one of the numbers being averaged? [yes]

B. Can the average be more than the largest number being averaged? [no]

C. Can the average be less than the smallest number being averaged? [no]

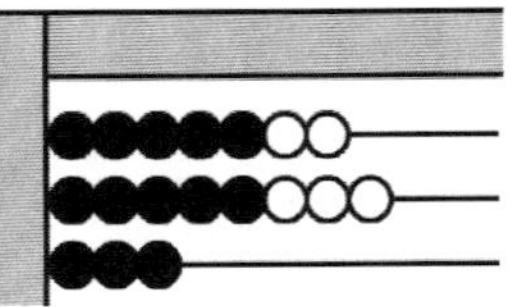 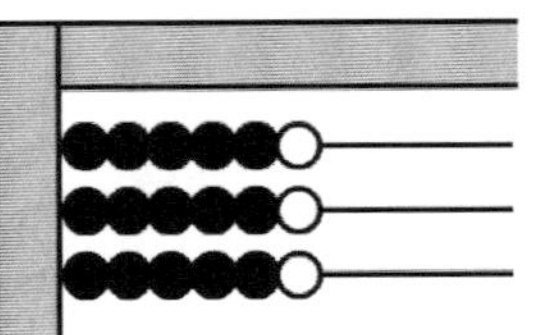

$$\frac{\begin{array}{r} 35 \\ 29 \\ 87 \\ +53 \\ \hline 204 \end{array}}{4} = 51$$

Find the average.

8, 7, 3 ____

2, 4, 8, 6 ____

9, 6 ____

7, 7, 7 ____

0, 5, 10, 6, 9 ____

81, 90, 72 ____

81, 90, 72, 72 ____

81, 90, 72, 0 ____

Unit 11
Japanese (Chinese) abacus

The Japanese abacus, with one bead above the beam and four below, is a very efficient calculating instrument for adding and subtracting. However, many hours of practice are needed to become an expert. In tests between a calculator and Japanese abacus, the abacus is faster for addition and subtraction, but slower for multiplication and division. The Chinese abacus is similar, but has larger beads and an extra bead both above and below the beam.

The extra beads of the Chinese abacus allow for an intermediate step in the carrying or borrowing step. The instructions given in this unit will not use the extra beads.

Learning the Japanese abacus is an excellent enrichment activity. Adults, as well as children, enjoy learning to use it.

Unfortunately, Japanese abacuses are very difficult to find in this country, but Chinese abacuses are available. With a few pieces of adhesive tape, the AL abacus can be made into a Japanese abacus. Use three narrow strips of tape, each long and wide. Place the abacus flat on a table in the vertical position. Slide the top five beads of each row to the top—these will not be used—and attach one piece of tape directly below them to keep them immobile. Fasten the tape on the top of the frame, down the side, across the wires, and to the other side. Slide the lowest four beads to the bottom and place the second strip above them, leaving a space equal to the height of a bead. Next slide the remaining single beads down to the tape and place the last strip of tape above them, again leaving a space. See the figure.

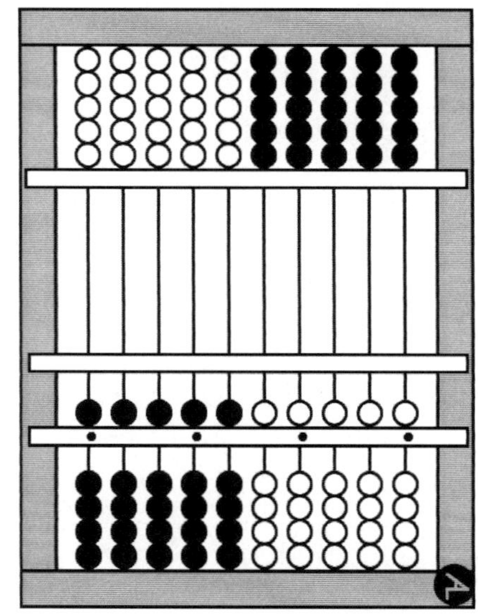

On the bottom strip, which now is called the beam, place dots at the first, fourth, seventh, and tenth wires, now called rods. (To start, you might prefer to place the dots at the second, fifth, and eighth wires in order to coordinate the ones with the colors.) These dots mark the units of the ones, thousands, millions, and billions.

Unless you are already familiar with the Japanese abacus, it is suggested that you practice the following exercises before teaching them. Worksheets for the children are available in the *Worksheets for the AL Abacus*.

NUMERATION

A bead below the beam has a value of 1 and is called a 1-bead; a bead above the beam has a value of 5 and called a 5-bead. The Chinese use the more poetic earth beads and heaven beads. A bead takes on value only when it touches the beam.

To clear the abacus, first tilt the upper edge so the beads fall to the bottom. Then clear the 5-beads by pushing them away from the beam. On an authentic Japanese abacus with its specially shaped beads, use the back of the fingernail on your index finger to clear the 5-beads.

The Japanese stress efficient manipulation for quick calculating. The following two rules are extremely important
and should be practiced from the start. Use only the
thumb for pushing up the 1-beads. Use the index finger
for pushing the 1-beads down and for pushing the 5-
beads both up and down.

1. On the ones rod, practice setting (the new word for entering) quantities 1 to 4 in one operation. For example, to
set 4, push up all four beads simultaneously. Next practice setting 5 to 9. For quantities 6 to 9, set the 1-beads
and the 5-bead in one "squeezing" motion. Children may
work in pairs setting numbers and reading them.

| 0 | 1 | 2 | 3 | 4 | 5 | 6 | 7 | 8 | 9 | 1 0 | 1 0 0 |

2. The next step is to set 10, 20...90. Then proceed to
numbers with tens and ones. The highest value is always
set first; that is, numbers are set in order from left to right.
Continue to larger numbers, even up to 99 million.

3. COUNTING TO 100. An excellent activity is counting
by 1s to 100. Counting to 4 is obvious, but to proceed to
5, push down the 5-bead and the four 1-beads in one con-
tinuous motion. To go from 9 to 10, the 1-beads and the
5-bead on the ones rod are cleared before pushing up a 1-
bead on the tens rod.

ADDITION

Adding without carrying

All operations are performed from left to right. Add the following.

33	22	11	11	55	66	33	55
11	22	22	33	11	22	55	44
44	44	33	44	66	88	88	99

12	36	12	56	51	53	48	72
76	53	35	31	45	46	50	27
88	89	47	87	96	99	98	99

341	210	453	516	1234	2550	7269
557	563	536	51	2150	2328	1730
898	773	989	567	3384	4878	8999

Carrying with the 5-bead

Think $4 = 5 - 1$. To add 4, push down the 5-bead and a 1-bead in one smooth operation. The shaded beads indicate the ones to be moved.

11	22	34	54	212	342	745	284
44	44	44	34	414	144	414	414
55	66	78	88	626	486	1159	698

Think $1 = 5 - 4$.

44	54	42	41	243	404	334	641
11	21	17	14	441	145	141	214
55	75	59	55	684	549	475	855

Now think $2 = 5 - 3$.

33	43	33	34	364	345	489	234
22	21	25	52	232	222	210	222
55	64	58	86	597	567	699	456

Here think $3 = 5 - 2$.

22	34	23	24	123	234	345	456
33	33	32	33	333	333	333	333
55	67	55	57	456	567	678	789

42	57	43	34	314	321	328	2601
36	32	43	51	243	434	431	4278
48	89	86	85	557	755	759	6879

Carrying with tens

When 5 is set, think $5 = -5 + 10$.

52	35	67	38	269	817	664	38
52	25	55	55	355	542	250	35
104	60	122	93	544	355	535	54
				1168	1714	1449	127

If 9 is set when 1 needs to be added, think 1 = –9 + 10.

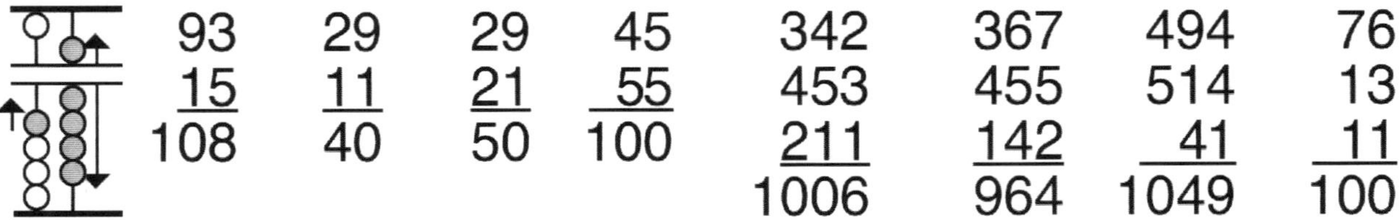

93	29	29	45	342	367	494	76
15	11	21	55	453	455	514	13
108	40	50	100	211	142	41	11
				1006	964	1049	100

Likewise, 2 may be thought of as 2 = -8 + 10.

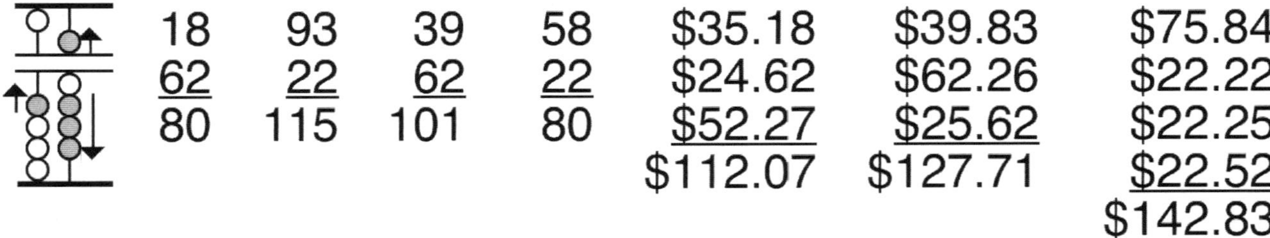

18	93	39	58	$35.18	$39.83	$75.84
62	22	62	22	$24.62	$62.26	$22.22
80	115	101	80	$52.27	$25.62	$22.25
				$112.07	$127.71	$22.52
						$142.83

Sometimes, 3 may need to be thought of as 3 = –7 + 10.

67	79	89	47	33 + 33 + 33 + 33 + 33 = 165
33	33	32	33	74 + 33 + 83 + 39 + 33 = 262
100	112	121	80	89 + 31 + 39 + 52 + 78 = 289

Almost half of the time 4 will be added by 4 = –6 + 10.

67	78	89	84	69	895	656	798
44	44	44	44	44	444	444	432
111	122	133	128	76	502	783	546
				44	444	444	234
				233	2285	2327	2010

Here is a fun practice row.

123,456,789	123,456,789	123,456,789	123,456,789
111,111,111	222,222,222	333,333,333	444,444,444
234,567,900	345,679,011	456,790,122	567,901,233

123,456,789	Add 123 ten times.
555,555,555	Add 234 ten times.
679,012,344	Add 345 ten times.

Adding 6, 7, and 8

Here adding 6 is thought of $6 = 5 + 1$.

55	16	27	308	286	867	758	6
56	66	66	606	365	246	606	6
111	82	93	914	664	766	655	6
				1315	1879	2019	6
							24

But sometimes $6 = -4 + 10$.

409	79	48	294	123,456,789	567,901,224
606	66	66	666	666,666,666	666,666,666
1015	145	114	960	790,123,455	1,234,567,890

One way to add 7 is $7 = 5 + 2$.

57	61	567	156	867	897	723	63
77	77	776	777	677	247	736	173
134	138	1343	933	723	843	176	6
				766	567	435	400
				3033	2554	2070	642

The other way to add 7 is $7 = -3 + 10$.

85	98	73	123,456,789	$7 + 7 + 7 + 7 = 28$
77	77	77	777,777,777	24¢ + 72¢ = 96¢
162	175	150	901,234,566	$73 + 237 = 310$

To add 8, one way is $8 = 5 + 3$.

55	56	16	66	$23.65	$10.67	$685.00
58	78	88	88	6.88	5.76	868.55
113	134	104	154	72.80	88.60	87.88
				$103.33	$105.03	$1,641.43

The second way is $8 = -2 + 10$.

25	234	789	541	123,456,789	34,567,801
88	888	888	878	888,888,888	88,888,888
113	1,122	1,677	1,419	1,012,345,677	123,456,689

Adding 9

The first way to add 9 is $9 = 5 + 4$.

55	50	65	567	867	78	405	3
59	99	89	967	806	234	95	4
114	149	154	1534	482	843	909	8
				99	699	571	9
				2254	1854	1980	24

The more common way is $9 = -1 + 10$.

11	234	789	567	123,456,789	23,456,789
99	999	999	999	999,999,999	99,999,999
110	1233	1788	1566	1,123,456,788	123,456,788

One final practice is to add 123, 456, 789 nine times and get the answer 1,111,111,101. This exercise uses most of the possible addition combinations. A "pro" can do this in 30 seconds.

For further instructions on the Japanese abacus, see *The Japanese Abacus: Its Use and Theory* by Takashi Kojima, published by the Charles Tuttle Co., Rutland, Vermont.

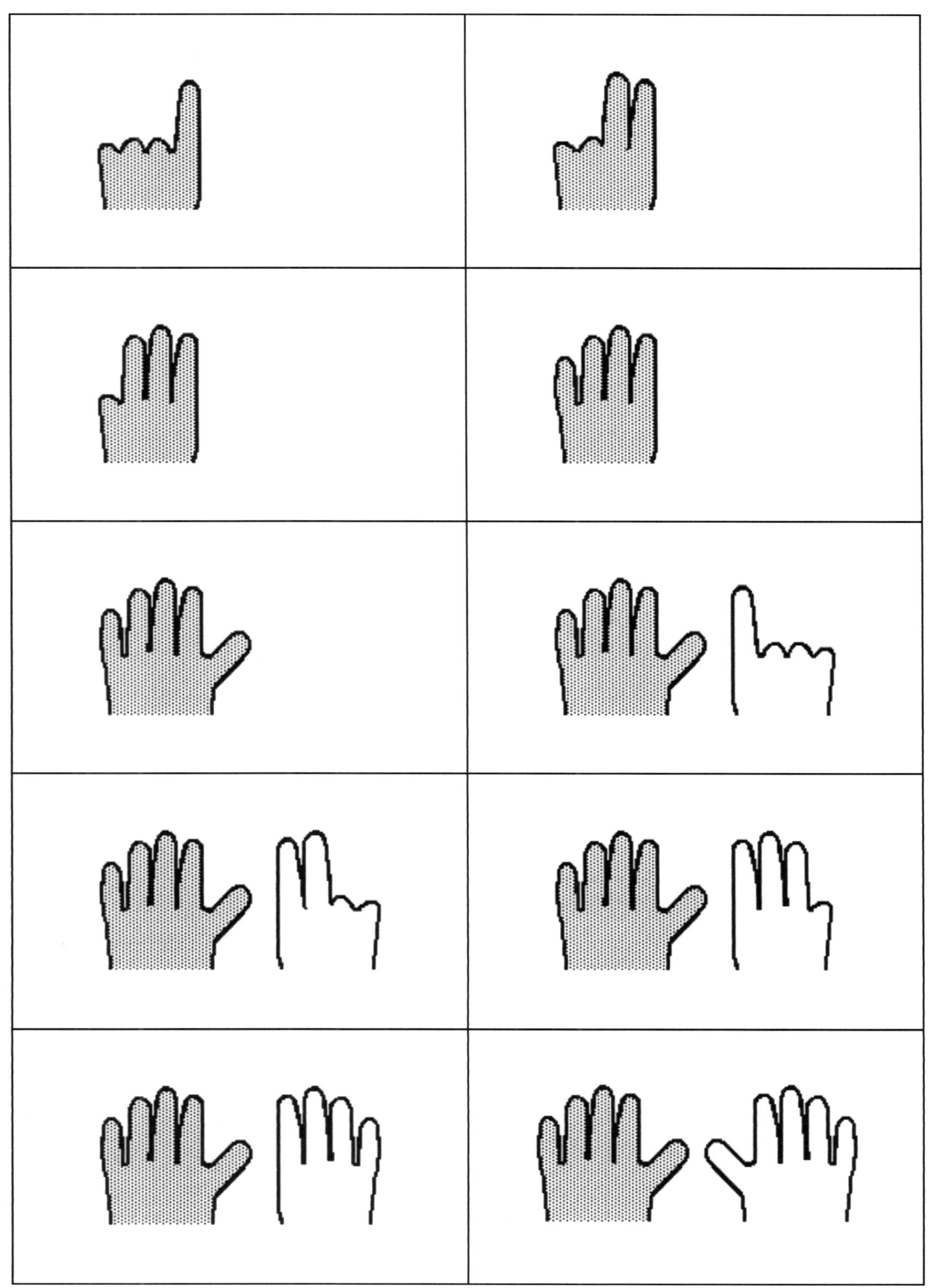

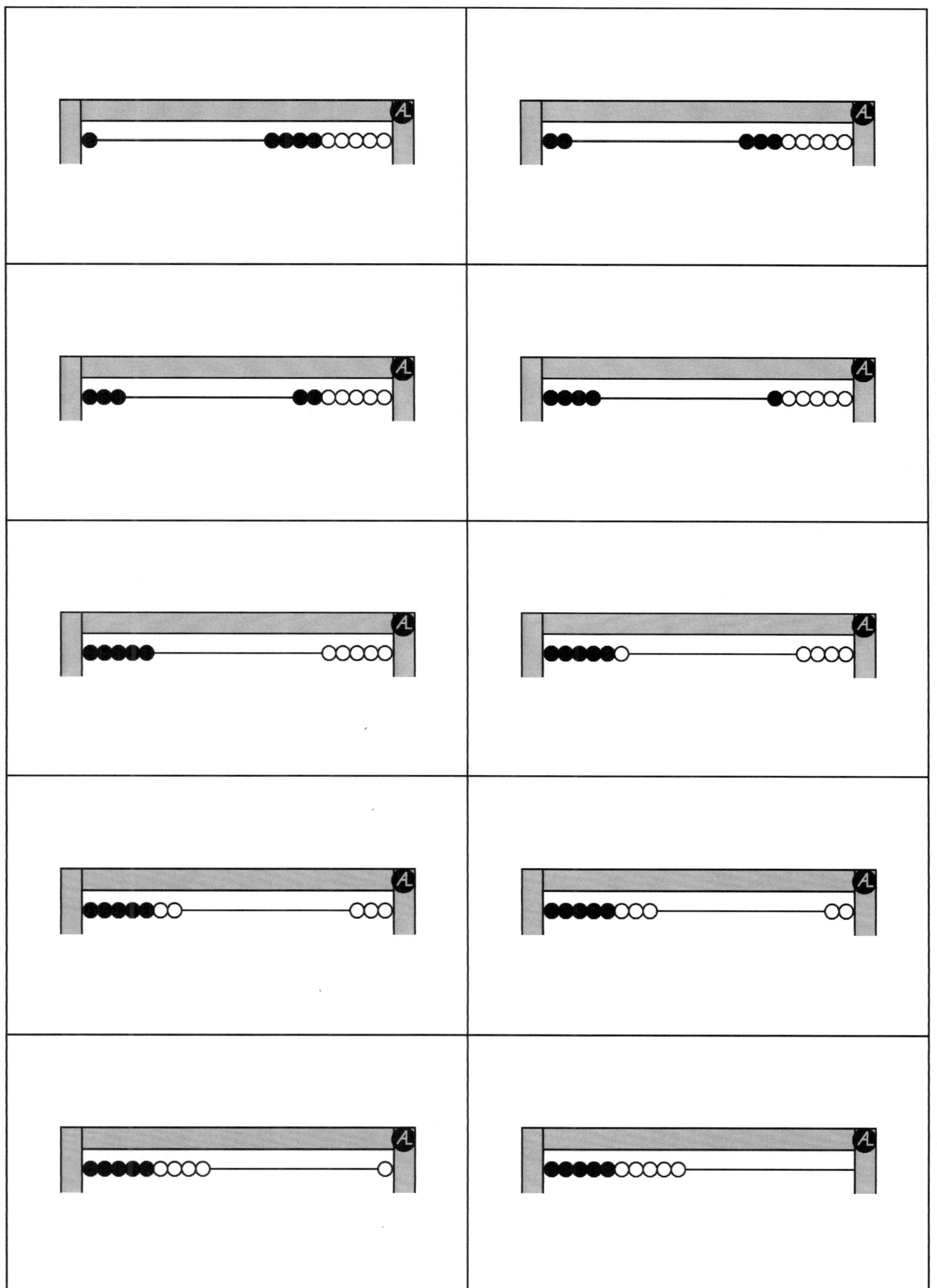

The numbers are backwards so they
can be traced (or glued) on the back
of the Safety Tread. A possible
layout is shown below on the left.

13.5 in

Placement of the red dots.

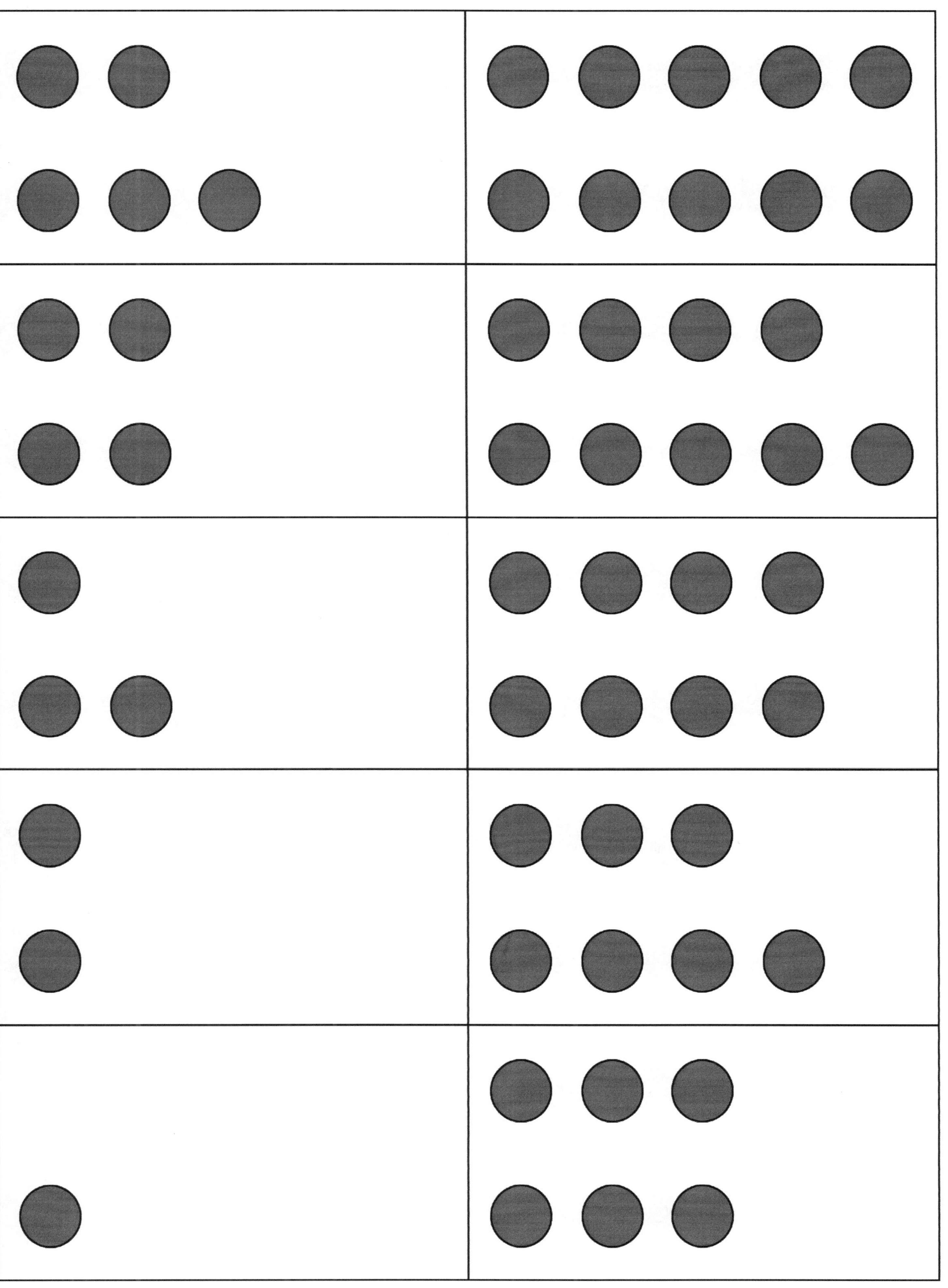

5	10
4	9
3	8
2	7
1	6

1 st	2 nd		
3 rd	4 th		
5 th	6 th		
7 th	8 th		
9 th	10 th		

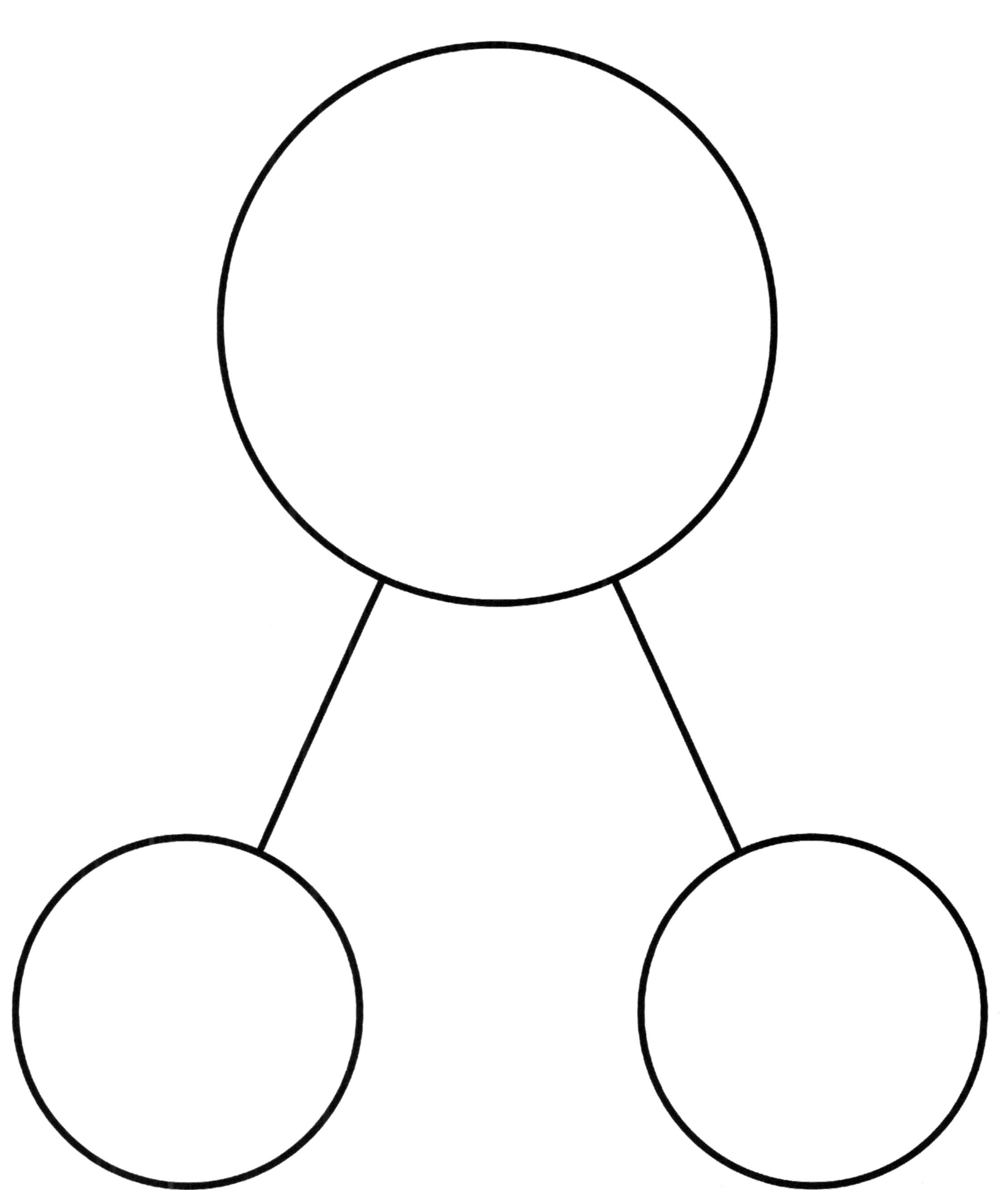

1 0 0 0	6 0 0 0
2 0 0 0	7 0 0 0
3 0 0 0	8 0 0 0
4 0 0 0	9 0 0 0
5 0 0 0	+ = −

1 2 3 4 5 6 7 8 9 10

Hundred Chart

1	2	3	4	5	6	7	8	9	10
2	4	6	8	10	12	14	16	18	20
3	6	9	12	15	18	21	24	27	30
4	8	12	16	20	24	28	32	36	40
5	10	15	20	25	30	35	40	45	50
6	12	18	24	30	36	42	48	54	60
7	14	21	28	35	42	49	56	63	70
8	16	24	32	40	48	56	64	72	80
9	18	27	36	45	54	63	72	81	90
10	20	30	40	50	60	70	80	90	100

Addition Chart

+	1	2	3	4	5	6	7	8	9
1	2	3	4	5	6	7	8	9	10
2	3	4	5	6	7	8	9	10	11
3	4	5	6	7	8	9	10	11	12
4	5	6	7	8	9	10	11	12	13
5	6	7	8	9	10	11	12	13	14
6	7	8	9	10	11	12	13	14	15
7	8	9	10	11	12	13	14	15	16
8	9	10	11	12	13	14	15	16	17
9	10	11	12	13	14	15	16	17	18

Cotter Sum Line

0 1 2 3 4 5 6 7 8 9

0 1 2 3 4 5 6 7 8 9 10 11 12 13 14 15 16 17 18

Subtraction Charts

- or	1 11	2 12	3 13	4 14	5 15	6 16	7 17	8 18	9	10
1		1	2	3	4	5	6	7	8	9
2	9		1	2	3	4	5	6	7	8
3	8	9		1	2	3	4	5	6	7
4	7	8	9		1	2	3	4	5	6
5	6	7	8	9		1	2	3	4	5
6	5	6	7	8	9		1	2	3	4
7	4	5	6	7	8	9		1	2	3
8	3	4	5	6	7	8	9		1	2
9	2	3	4	5	6	7	8	9		1

	1	2	3	4	5	6	7	8	9	10	11	12	13	14	15	16	17	18
−1	0	1	2	3	4	5	6	7	8	9								
−2		0	1	2	3	4	5	6	7	8	9							
−3			0	1	2	3	4	5	6	7	8	9						
−4				0	1	2	3	4	5	6	7	8	9					
−5					0	1	2	3	4	5	6	7	8	9				
−6						0	1	2	3	4	5	6	7	8	9			
−7							0	1	2	3	4	5	6	7	8	9		
−8								0	1	2	3	4	5	6	7	8	9	
−9									0	1	2	3	4	5	6	7	8	9

Multiplication Table

x	1	2	3	4	5	6	7	8	9	10
1	1	2	3	4	5	6	7	8	9	10
2	2	4	6	8	10	12	14	16	18	20
3	3	6	9	12	15	18	21	24	27	30
4	4	8	12	16	20	24	28	32	36	40
5	5	10	15	20	25	30	35	40	45	50
6	6	12	18	24	30	36	42	48	54	60
7	7	14	21	28	35	42	49	56	63	70
8	8	16	24	32	40	48	56	64	72	80
9	9	18	27	36	45	54	63	72	81	90
10	10	20	30	40	50	60	70	80	90	100

1. Carefully cut around the outline.
2. Fold on the heavy lines.
3. Assemble the cube with the flaps on the outside. No glue or tape is necessary.
4. If the flaps do not lie flat, pinch gently on the fold line.

Abacus Tiles

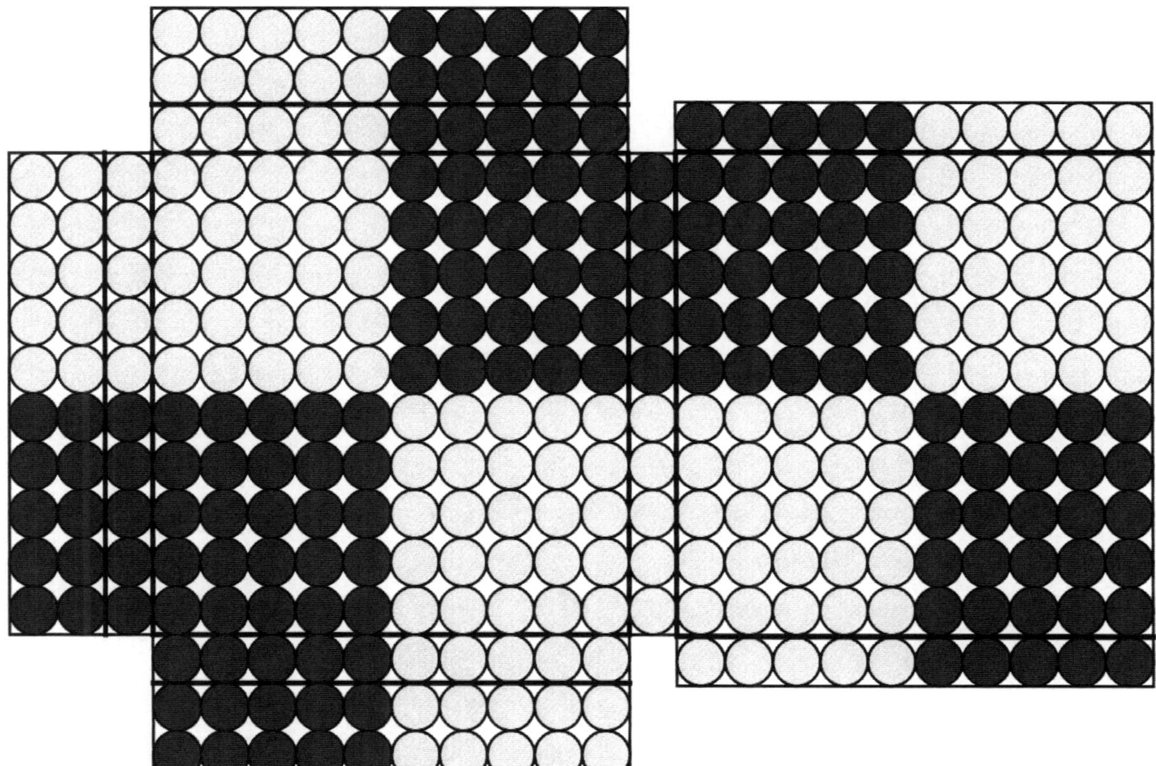

1. Copy on heavier paper (card stock).
2. Cut around the outline.
3. Fold on the heavy lines with the printed side out.
4. Assemble into an abacus tile, as shown on the right.
5. Tape or glue the flaps.

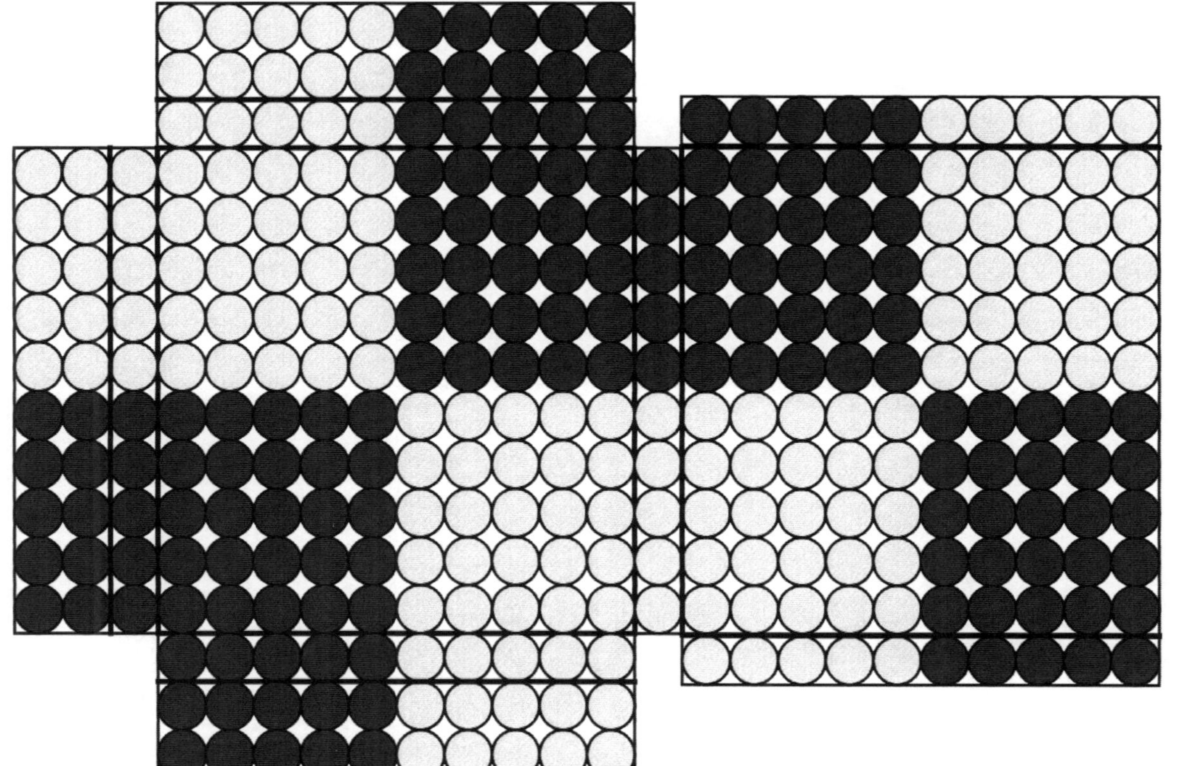

Note: Abacus tiles are available separately.